Computational Social Sciences

Computational Social Sciences

A series of authored and edited monographs that utilize quantitative and computational methods to model, analyze and interpret large-scale social phenomena. Titles within the series contain methods and practices that test and develop theories of complex social processes through bottom-up modeling of social interactions. Of particular interest is the study of the coevolution of modern communication technology and social behavior and norms, in connection with emerging issues such as trust, risk, security and privacy in novel socio-technical environments.

Computational Social Sciences is explicitly transdisciplinary: quantitative methods from fields such as dynamical systems, artificial intelligence, network theory, agent based modeling, and statistical mechanics are invoked and combined with state-of the-art mining and analysis of large data sets to help us understand social agents, their interactions on and offline, and the effect of these interactions at the macro level. Topics include, but are not limited to social networks and media, dynamics of opinions, cultures and conflicts, socio-technical coevolution and social psychology. Computational Social Sciences will also publish monographs and selected edited contributions from specialized conferences and workshops specifically aimed at communicating new findings to a large transdisciplinary audience. A fundamental goal of the series is to provide a single forum within which commonalities and differences in the workings of this field may be discerned, hence leading to deeper insight and understanding.

More information about this series at http://www.springer.com/series/11784

Anamaria Berea

Emergence of Communication in Socio-Biological Networks

Anamaria Berea
Robert H. Smith School of Business
University of Maryland
College Park, MD, USA

ISSN 2509-9574 ISSN 2509-9582 (electronic)
Computational Social Sciences
ISBN 978-3-319-87821-8 ISBN 978-3-319-64565-0 (eBook)
https://doi.org/10.1007/978-3-319-64565-0

Printed on acid-free paper

This Springer imprint is published by Springer Nature
The registered company is Springer International Publishing AG
The registered company address is: Gewerbestrasse 11, 6330 Cham, Switzerland

Acknowledgments

This work would not have been possible without the support of a grant from the Keck Futures Initiative of the National Academies of Sciences, Engineering, and Medicine, and I will be always thankful to them for opening this path in my career. I would also like to thank Professor James Kitts for supporting this book proposal, my graduate students Marta Hansen Marksdattur and Ruchika Satalkar for their extraordinary help with programming and literature review, Arun Kaipuzha for great insights into biological science, Professors William Rand and Roland Rust for the support of the Center for Complexity in Business in hosting this grant and allowing me to pursue my research passions, and my husband for his unconditional support of my research career. I would also like to thank my PhD advisor, Maksim Tsvetovat, for supporting this idea that started as a paper assignment in the social network analysis class. Last but not least, I would like to thank my editors at Springer Nature, especially Chris Coughlin, for their patience with my writing rhythm and supporting this project by allowing me the necessary deadline extensions.

Contents

Chapter 1
Introduction

The unifying theme of this book is that communication is an underlying fabric of life, as fundamental as matter and energy are to our world and, more importantly, to our understanding of the world.

This book is also a non-exhaustive account of interesting, out of pattern communication and social behaviors that we can observe in animals, or in the biological world and among us, humans, and between us humans and other species.

While we know that communication is fundamental to life, in any form or shape it comes on our planet, the goal of this book is to show that there are aspects of communication that can be universal and transferable from one species to another and that it is what enables collective behavior in social animals and humans. At the same time, there are aspects of communication that are unique to each species or ecosystem, that communication has evolved both alongside the genetic evolution, the social evolution that is characteristic to a subset of living species, and the cultural evolution that is characteristic not only to humans, but also to whales, dolphins, primates, elephants, and many more.

Moreover, this book points towards the need of a more fundamental theory of communication. In order to understand the future of languages, the development of artificial intelligence and perhaps even the development of interspecies communication, we need to pay attention to the fundamentals of communication in a way that is usable and transferable into practice, by trying to dance around tautologies and ubiquitousness of this phenomenon that is so universal, yet so poorly described in universal terms, and that is so fundamental, yet lacking a robust fundamental theory or fundamental platform of research.

Many researchers are drawn to exploring and researching one of the hardest questions in science: which are the origins of human language? I am not looking at this question here. This project started with some seemingly simple set of questions: what is the difference between the communication embodied in silicone, in our inanimate computers, and communication that is embodied in carbon, in our

A. Berea, *Emergence of Communication in Socio-Biological Networks*,
Computational Social Sciences, https://doi.org/10.1007/978-3-319-64565-0_1

living beings? Why is it so hard for species to transcend the boundaries of their bio-communication niches? And what happens if we relax the original theory of information exchange from Shannon?

While it is often understood and implied that information is the third foundation building block of our Universe, alongside matter and energy, and we intuitively grasp that information is an underlying fabric of the very existence of the living world, we don't have a clear understanding to day on how communication emerged in the living world, whether it is an intrinsic property of the living and not one of the non-living (although the "silicone" world may prove us otherwise), how it relates to social and collective behaviors and how ultimately influences our survivability, adaptability, and growth.

Additionally, most studies in communication are predominantly focused on the business or personal communication—therefore predominantly focused on the human communication in our days, at the peak of civilization. Also, some studies point communication towards psychology, others look exclusively at animal communication, species by species. And other studies loot at software engineering and IT as they place communication between computers or between humans and computers at the core of disciplines such as data science and cybersecurity. There is also an entire field dedicated to human-computer interactions and virtually any field has some studies or aspects of communication thoroughly embedded in it. But to my knowledge, an extensive study on communication per se, that transcends disciplines, species, means of communication, ways of interpretation and time has not been done before.

The evolutionary paradigm seems to be the current dominant scientific paradigm of the living systems, from biology to sociology. The relativistic/mechanistic paradigm seems to be the larger, non-living physical systems current paradigm in science. In human systems (social sciences and economics), there is a shift towards the less orthodox, mechanistic, and more heterodox, evolutionary approaches (more towards the "smaller" paradigm of evolution than the "bigger" paradigm of the universe). And most science today seems to be the science of the niches—incremental approaches of sub-sub phenomena that fall under either of the paradigms. Is this the end of Big Science? Have most of the important laws and theories in science already been discovered? And if so, how do the evolutionary paradigm of the living and the mechanistic paradigm of the non-living coexist? If we are a bag of atoms, we fall thanks to gravity; if we are a bag a cells, we mutate thanks to evolution; but if we are a bag of information, do we obey the laws of evolution, the laws of physics or none? Ultimately, doesn't any system (of cells or atoms) obey the laws of the other system they are in relationship with (humans in relationship with the Earth obey gravity; Earth in relationship with ecology evolution; ecology in relationship with humans obeys economics)? And wouldn't this mean that the scientific paradigms are the paradigms describing the relationships between complex systems and not paradigms describing the systems themselves? And what sciences study complex relationships between systems better than network science and economics?

In a way, all there is, can be described by matter and energy. And when the matter and energy become alive, it evolves. But what binds together matter and energy is information. The information contained in the atom of Carbon makes it bind with the information contained in the atom of Oxygen. It is the information contained in the matter and the energy that describes an atom or a molecule that makes any relationships between any atom and molecule with other systems possible or impossible. It is the information contained in each of these systems the one that draws the line between infinite possibilities and a finite world, that draws the line between determinism and uncertainty. And in a way, we can view this information, that is binding and bounding, as the communication that takes places between two entities or systems. This communication is the one that encrypts the possible from the improbable and can separate the existence from the non-existence. It is this communication that I am interested in finding how it emerges, how it is created and why, why it does not form and why it cannot exist under most circumstances. Particularly in the living systems, more or less social, it is this type of communication that is responsible for collective behavior, for the super-organisms and ultimately for our civilization.

It is this type of communication emergence that makes our world possible and all other worlds probable.

Just like communication and information, emergence is a very tricky word, with multiple meanings and understandings, from the broad meaning of revelation of something that has been concealed to the more restrictive meaning from systems theory of a pattern that one can observe in a system that is not observable or reproducible in the parts of the systems. It is a property of a system that reveals itself from the small interactions of the parts. Emergence is a property of a collective that becomes or reveals itself to be something else or more than the sum of its parts.

Emergence thus has some mystical connotations attached to it: we act and interact locally and something greater and bigger than us, with little predictability and determinism, reveals to us. The word itself is based on the Latin "emergere," to show up, and in the Middle Ages it was used with the meaning of "unexpected occurrence." It is also interesting to understand how the words of emergence and emergency are actually closely connected. Coming from the same Latin root and similar understanding, they imply the sudden revelation of a process or an event that has deeper roots and took a much longer time to "brew" out of sight. It is a process of self-actualization of agents and interactions that has been happening for a while. Nothing happens over night, only our understanding and awareness is surprised by patterns that seem to become entities of their own.

Chapter 2
Economics of Information Meets Social Networks: From Hayek's Mental Maps to Mathematical Linguistics

2.1 Why Economics of Information?

At this point in time, the economics of information already has a history of more than half a century of theoretical and practical developments. Yet there are still some challenges that the field is facing with respect to establishing a concentrated, focused theory of economics of information. Perhaps this is due to the fact that information is "everywhere," is so incredibly ubiquitous, and can sustain so many definitions and so many forms under various contexts, subfields, and analyses. Most of the work in the economics of information pertains to the analysis of information for economies, for management and marketing, industrial organization and policy implementations, all in the context of markets or institutions. But there is a small body of research that looks in more depth at information as an economic good and at some of its fundamental properties and behaviors as an economic good [14], and that is what economics of information currently is and mostly is understood as.

But economics, in general, is a science of human behavior. Where does economic behavior start though? Could we consider the evolution of life forms as one type of behavior, and by extension as a sort of biological/social/economic behavior? Could we, for example, start by saying that life forms have evolved from biological to social and only afterwards to economic beings? Can we consider all economic behavior as a subtype of social behavior or is there economic behavior that is not social? Can we consider all social behavior as biological behavior too?

All these questions are important to consider if we truly are to use a methodological reversal, as this book is attempting to do (Fig. 3.4). And by methodological reversal I mean applying theories and models from the economics and social sciences into biology or even the physical world.

What is the difference in analysis and understanding when we move from framing our place of departure from studying a very large spectrum of communication into framing our place of departure to study a continuum of behaviors and behavioral processes? For example, can we say that social behavior started at *X* point on

A. Berea, *Emergence of Communication in Socio-Biological Networks*,
Computational Social Sciences, https://doi.org/10.1007/978-3-319-64565-0_2

the evolutionary scale and that economic behavior started at $X + t$ point on the evolutionary scale? And why does this matter for the subject of this book, of the emergence of communication?

Economic behavior and the development of language or the development of forms of communication are closely linked. For example, when two different groups that don't speak the same language meet (think about the first encounters between the Spanish soldiers and the indigenous populations of the Americas or between the first traders along the Silk Road)—body language and the exchange of objects connects the different groups in a friendly, cooperative way before the exchange of words and the possibility of translation between the two languages can begin.

On another hand, business research also shows that people that communicate more are more likely to engage more in economic behavior with one another, from employment to business contracts. Communication that leads to more communication is a signal of trust and cooperative, harmless behavior, while communication that leads to less and less communication is a signal of mistrust or non-interest in any communication behavior [34, 77].

Another question that we need to address is whether economic behavior is a subtype or subset of social behavior. Is all economic behavior social and is there social behavior that is not necessarily economic? Given that social behavior is found in many species, but economic behavior only in a few, we would assume so (Some people say that economics is the "queen of social sciences"), but a deeper examination and research is required in this respect. We intuitively know that there is a very vast universe of social behavior that is not necessarily economic: going to parties, romantic relationships, playing amateur team sports, and many more. On another hand, there is a lot of research that tries to explain even the most mundane social behavior through the lens of economic thinking, particularly through the lens of value theory and/or signaling theory.

Much of human economic behavior relies on assumptions about signaling, exchange, perfect or asymmetric information; on market coordination or mis-coordination; on assumptions about conflict or cooperation. All these assumptions—or variations on assumptions—in the multitude of economic models ultimately relate in one way or another to the communication between humans or human institutions.

But what about the animals? Do they also engage in economic behavior and hence economic type of communication, or do they only exhibit various forms of social behavior (at least some species that we know are social)? I am exploring some of these questions in this book and model that I am proposing here.

Hayek's "Sensory Order" [39] is a very unique and intriguing book written by the Nobel prize economist with the purpose to explore the roots of economic spontaneous order in the human mind. But what the book also achieves is the explanation of the sensory experiences based on the network of brain cells in humans.

Although Hayek's work is not unusual or uncommon in medicine and psychology, and we know that the brain is a network of cells and that the representation of information and memory is also based on associations of concepts that evolve in time, Hayek's contribution is important also because he is a well-known economist of information. In his most influential work, he has shown how prices are the mechanisms to convey the most relevant information on the market [40]. Therefore he is an author that has practically unified the ideas of information and networks both in society and in the human brain.

Hayek's contribution to our understanding of the mind and of language was, at the time, an attempt to counteract behaviorism as a way to explain economic behavior as being purely subjective, sensorial responses to the environment. At the same time this is an attempt to rather try to understand the mind rather than to predict behavior through a set of principles, similarly to how social phenomena is better understood than predicted, as well as to explain the roots of human action through its own representation [17].

Hayek also describes that selection, evaluation and interpretation, as well as learning are parts of the sensory order, therefore introducing the idea of selection and "meaning"—the interpretation our brains give to any human action. These ideas would be later on widely accepted in science, but at the time they served the purpose of understanding human economic action. In this way, he basically laid the foundations of representing the mind as a complex system with emergent, evolving spontaneous order, that could not be fully predicted, but for which we can detect patterns, similarly as we do in social science to describe how we view and understand society today.

2.2 Signals vs. Information vs. Networks

Communication is arguably one of the largest and most interdisciplinary fields in science and social science today. There is a difference between communication and information though, that makes the study of communication in social science, information science or economics particularly interesting.

In order to understand communication, we need to understand the basics of information and information exchange. And in order to understand information exchange, we need to understand its characteristics and behavior, as a phenomenon or as an agent that drives a phenomenon. But how do we model information? How do we represent information so that we make sense of its role in biology and social systems?

The fact that information in living systems is best represented as networks seems intuitive—the neurons and the brains of the vertebrate beings show physical interconnectedness and we now know they also display information interconnectedness. Perhaps our best models of the physical representation of the information that we intuitively think about today are the brain and the Internet. The brain is formed of a network of neurons and the Internet is formed of a network of servers and computers that are linked through cables spanning the world (including the oceans). But this does not necessarily mean that a network representation of information would be the best model that we could explore in the scientific research of communication. Since the evidence about how information is transmitted, exchanged, produced, and consumed by any biological organism shows that its dynamics is neither linear nor uniform, currently network representations of information and communication are the best paradigms for modeling that we have at our disposal.

There are several formal ways to represent information and other several ways to model information, but my argument in this book is that some models of network or graph representations are better than others to model and understand information. Let's start with some crude examples first. When we think of information, our first thoughts that are likely representing in our mind are probably images of the human brain or the computers (again, either cognitive or digital). Information is a large enough and abstract enough concept so that our immediate physiological response would be to associate it or mentally represent it as the most common image we see or remember, and this image is more often than not the one of a brain or the one of a computer. We hardly think of information when we observe a piece of art or nature although both are also information, just in different forms.

At an interdisciplinary conference organized in 2014 by the National Academies of Sciences Keck Futures Initiative, we debated the topic "Collective Behavior—From Cells to Societies" [87]. At this conference, we tackled some fundamental questions about how collective behavior emerges, and how the living and the social worlds scale from 1 entity to 2 (pairwise interactions) and from 2 to more. If we are to accept the assumption that cells and societies have similar processes of growth, essentially that there is some fundamental process that leads to collective behavior both in biological and in social systems, then what we should observe out of this scalability or growth are similar structures or results—similar shapes or similar topologies or similar statistical regularities arisen from the same process of growth.

Living organisms are not Poisson distributed (or, in other words, there is no evidence of random networks in living systems), which means that there is a "meaningful" relationship between any 2 organisms of the same species on this Earth. Also, there is little evidence of duality (only 2 individuals) in biology or sociality... there is only "many." We don't find only 1 or 2 instances of something in the living world; neither 1 or 2 individuals of any living organism could be sustainable or exist in the long run of epochs or eons. Yet 2 is the link that bridges one with many. In information theory, we have mostly 2 as a building block: a sender and a receiver, and the binary representation of information through 1s and 0s. In biology, information is represented by 4 or 5 or 6—the "many" building block (DNA molecules). But also in biology, morphogenesis is represented by 2 as an originary building block: there is a cell that divides first in 2, and then in 4, and then so on. Division and multiplication are binary, but behavioral aggregation is "many" merging or fusing into a new entity. As processes, they go into different directions (one to many vs. many to one). On another hand, in network science, the original elementary "structures" are formed by 3 (triads) and 2 (dyads). In social systems, information is only represented by "many or multiples." There is no social group with only 2 elements.

This numerical symbolism, which may seem drawn from the semiotics of the Middle Ages, is actually important in order to grasp the differences between communication that emerged in biology through millions of years of evolution versus communication that we designed during the last few decades—natural versus artifactual communication. The scalability and evolution of the natural communication are much harder to grasp and understand than the scalability and evolution of the digital communication during our still fresh recent history.

One thing that certainly sets apart one type of communication versus the other: the Poisson distributions of the artifactual communication, from the telephone to the digital life, in the realm of computer communications [28]. Mathematically, random networks exhibit a Poisson topology [27]. This means that "silicone" or artifactual communication is perhaps more likely to exhibit a Poisson distribution, while "carbon," living organisms type of communication is not—they are more likely to cluster in "small worlds"[93]. The meta-networks or the global networks of communication are more likely to be Poisson and random, while the local networks or individual networks are more likely to be small worlds. The "human brain connectome," which represents the network topology of neural connections in the brain, shows that the human brain has a complex topology, with hubs and clusters, and modularity, which allows the brain to be both efficiently interconnected and to be able to function if large parts of the brain connectivity are lost. Mapped as a graph, all statistical analyses on the human connectome data showed that the brain has small-world networks characteristics [32, 37].

While Hayek showed the connectedness of information in the brain and in the market, Brian Skyrms "Signals" [81] tackles the problem of signaling from a game-theoretic perspective. In this book, the author is presenting several game-theoretic models of signaling, that can be applied to both animal communication and in economics, to signaling and asymmetric type of behavior for humans under various economic circumstances.

John Holland took a different approach to signals; he has been one of the most exquisite complex systems and complexity science scholars, whose books, even if written for a wide audience, have made their way into the classroom textbooks. Particularly Holland's book, "Signals and Boundaries: Building Blocks for Complex Adaptive Systems," is foundational for understanding communication from a network science and complex systems perspective. His thesis here is that any agents, whether they are molecules or humans, "signal" their actions when they interact with each other, but this signaling is bounded by the meaning of the boundaries or roles that the agent is acting in. For example, the molecule can interact with other molecules, being bounded by the organism that it lives within, or a person can interact with another person based on the role they are acting in (i.e., parent–child versus professional roles versus friendship roles) [43]. While these boundaries are also fluid and they change and evolve, in a sense they predetermine the meta-level of information where the actual information exchange is taking place.

The complex systems of signals and boundaries can also be seen as hierarchical systems where individuals form collectives and interact with each other within the collectives. In other words, he addresses the problems of scale and of collective emergence from the multitude of signals that can only be produced and received within a certain context or hierarchy and not in the abstract or free form.

These very important ideas outlined above are overlapping with some other ideas from economics, philosophy, game theory, animal communication, and many more, regarding signal s and communication.

More than just a thesis for a new understanding of signaling, communication and language emergence, this book is also a methodological manifest for exploring computational approaches as viable alternatives to theories in science and the overlapping theories and models from economics, information theory, network science and complex systems will help shape a better, scalable view of communication in life forms and digital artifacts.

2.3 Economics and Language: Human Specific Phenomena?

The link between economics and language has been studied for a long time. There is even an entire field of study on economics of language, that is looking at the effects of language or language gaps in the economy, as well as the influence of a global language on the global economy.

But economics also provides a very useful way of thinking to study, model, and understand the phenomenon of information. As I mentioned before, the Nobel laureate F.A. Hayek is one of the prominent economists that makes the link between information in economies manifested as prices and the information in the brain manifested as "mental maps." His particular position in understanding information exchange comes from his background and interest in psychology and cognitive processes as well as economics.

Economics is not only a science that aims to explain economic growth and alleviate inequality, reduce poverty but it also has great insights into human behavior and human forms of organization and the origins of our civilization. Economics is an incredibly powerful "tool" for understanding human behavior, particularly collective human organizing behavior that is at the core of the evolution of our civilization. Communication, language, and art are products of our civilization as well as representations of our current stage in civilization. Economics emerged in our lives when humans had to organize themselves to cultivate, not just to harvest resources (agriculture) and also to defend themselves against members of the same species (conflicts). Economics coevolved with the technological improvements and the current developments in languages and information would not have occurred unless we had an economic system. Economics taught us the coevolution of civilization and natural resources, with both good and bad outcomes. The differences between the good and the bad outcomes come from our continuous experiments with the design of our institutions and rules of economic behavior. The rest is embedded in our natural world.

Economics and language found a common ground in political economy and the science of governance as well. Vincent and Elinor Ostrom, two famous political economists (Elinor being the first woman to receive the Nobel prize in economics) have studied all their lives the problem of governance, the interplay between the individual and the collective behavior, between managing private and common resources and the best ways for people to self-govern[63]. The transition between individual and collective behavior in political economy has also led Professor Ostrom to discuss language and culture in the context of markets and the exchange of ideas [62].

In the context of societies, language is the key ingredient for creating social order and decision rules. Language also plays a fundamental role in the shared knowledge and the shared common understanding of humans. In the long span of history, shared language communities have become countries and nationalities and thus Ostrom ties the shared language and communication in humans to orders of collective behavior [62].

Ostrom identifies language as the primary way to gain access to stored knowledge and for information to flow on an intergenerational basis in humans. A similar argument is made by Whitehead and Rendel in the animal world, where they identify cultures in cetaceans [94]. Ostrom also asserts that "communication happens in shared communities of understanding" and that the syntax and semantics (and, in general, the grammar) are what separates us from other organisms—through these we can convey "meaning". I find very interesting the fact that the idea of "meaning" has emerged at the intersection of 3 coevolutionary processes in humans: biological(cognitive), collective/social behavior, and grammar.

He also asserts that human languages "are crafted in concurrent patterns of communication in associated relationships that are constitutive of ways of life among contemporaries" [62]. This means that even from the political economy and governance of the collectives point of view, communication and languages are an underlying fabric of life. He goes on by mentioning that contemporaries create artifacts that contribute to the intergenerational cycle of life. As Herbert Simon also alluded in the "Sciences of the Artificial," artifacts play a significant role in the coevolution of our civilization and our environment, and in the emergence of languages and the recording of language for intergenerational transmission of knowledge and culture [79].

Also within the political economy framework of thinking, Ostrom warns against the misuses and abuses of language, particularly through a superabundance of language that would only increase dangers of biases and false beliefs, as well as of production potential, while a scarcity of language is desirable—in other words, an efficient language is more desirable than a language that only creates a larger and larger search space for meaning. Therefore selection and adaptation processes should be taken into account when studying language.

Economics and language intersect not only as economic reasoning and economic applications applied to language or as philosophical constructs for social behavior, but also in a more applied, daily and contemporary way, such as the effects of language into family economics or macroeconomics [7]. In this subfield of development economics, the researchers are trying to understand the effects of language barriers into the economics of communities, migration economics or the global economy at large. From the perspective of this book, the central idea to highlight from these studies is that language or communication barriers create new patterns of social behavior and that global, more uniform language and communication patterns create more global and uniform social behavior.

Which are the constraints that shape language and communication? Which are the boundaries that can determine the form and function from noises into a meaningful way? Marvin Minsky asserted that one of the universal features of language and communication in the universe is that it is subject to the same physical constraints of space, time, and matter as life itself is [59]. And what better science to study the constraints or resource limitations than economics?

2.4 Communication and Coordination: Social Behavior in Living Organisms

Collective behavior is inherently the result of social behavior when individuals communicate and coordinate with others for both an individual and a common goal. Collective behavior is not specific only to humans, its roots go all the way to cellular and microorganism communication and coordination. Coordination in social groups, whether animal or human, is unachievable without communication. Similarly, communication in social groups at the individual or group levels would not be possible without coordination; although in humans, what distinguishes us from the other species is the fact that we can delay coordination for communication purposes. This "delay" and "storage" of communication, through the means of writing or other types (papyri, manuscript, book writing, and later on emails and digital storage of collective work and tasks (such as Dropbox, Jira, and so on) achieve just this), we can coordinate and communicate more through these means that facilitate communication while we are not physically present. In the animal world, storage or delay in communication is much more rare.

In the animal world, there is delayed communication through chemical means (pheromones, scents) or tracking the footprints of prey, as well as there is some evidence of intergenerational transmission of culture and use of tools in cetaceans [94]. But innovations in means of communication that would facilitate wide transmission and storage of valuable communication (such as writing) is not present in the animal world.

But, overall, communication and coordination are so intrinsic and co-dependant on each other that it is hard to distinguish one from another. Miller and Moser proved using game theory that communication leads to superior outcomes in coordination [58]. More than that, they showed that the original communication or information exchanged could be meaningless. The exact mechanism through which communication enables superior coordination outcomes is not clear, but what both game theory and agent-based simulation models show is that communication enables the emergence of "an ecology of behaviors" [58] that leads to more information and better coordination.

And just as in the applied model presented in Chapter 4, the authors mention that whether we are looking at molecules, organisms, or economic entities, the ability to coordinate is a key feature of the social world.

Intuitively, we think that of course there is a strong link between communication and coordination. Yet this link is understudied. Coordination is essentially a problem of time synchronization. More than the aspect of time, coordination involves not only the actors to be situated in the same period of time and in the same space (less so for humans in the modern world), but to also "synchronize" with each other. Many researchers and scientists agree that problems of coordination are as fundamental and as fuzzily understood as are the problems of scalability. The systems in our world, either computational, biological, social or institutional, can be modeled and understood only as slices of bigger systems in slices of predetermined periods of time. In another way to understand this, scalability is a problem of ontological space, while coordination is a problem of syntax or grammar or rules that keeps this ontological space together.

But when you change the scale (to a much bigger or a much smaller system) or when you change time, the same models and similar tested behavior break down. For example, if you think of the international financial system, or the international trade, there is no central governance that rules over the actions of the people or institutions or countries engaged in this behavior. On another hand, the financial system or the trade system of a country, which is part of the larger international scheme of things, is being governed by a different set of rules. And as we "go down" the scale, a market in a city functions very differently than the system of the country. In biology, specific ecosystems are being ruled and governed by principles and laws that may or may not be found in smaller systems—the fauna and flora of the mountainous areas are different based on various altitudes (Mount Kilimanjaro is a great example in this respect: at the same altitude, the fauna and flora differ in different geographic locations; another example is the difference in climate and ecosystems between the Andes and the Alps), while even the same species exhibit different behavior based on where or when you find it (some good examples would be the whales in the Pacific versus the whales in the Atlantic).

Coordination and scalability are not meant to confuse us though. While nobody really knows which are some "good boundaries" for selecting the proper scale and coordination mechanisms of a phenomenon, we are still modeling it and finding interesting results. A lot of the "good" scale and coordination selections in research or our understanding of the world has to do with our intrinsic perceptions and habituation of the world we live in—therefore they are relative to our understanding and perception of the world.

Coordination is not only a problem of time and space (the actors would need to be in the same place at the same time), but also of intent and of possibility of coordination. Actors in the same space and time can still mis-coordinate. Here is where communication comes into picture—based on the interpretation of signals and responses, the actors react or intentionally decide to coordinate with each other or not. Communication is therefore that "glue" that helps coordination happen or not, when all other conditions are fulfilled.

Chapter 3
Heterogeneity and Subjectivism in Communication: From Trust and Reputation Mechanisms to Linguistic Norms and Collective Behavior

3.1 Collective Behavior and Communication

In a very simplistic way, we can assert that communication is a complex adaptive system because the organisms that communicate are complex adaptive systems themselves. This seems tautologic, but since communication is all about the means of transmitting and receiving the message and about the message itself, it heavily involves the sender and the receiver in ways that other types of exchanges do not (such as the exchange of goods or services). The means by which the sender and the receiver get involved in the exchange are also information and this makes information exchange incorporate both information and meta-information, thus rendering an epistemic and ontological relationship that is different from the one of the exchange of goods and services.

In animals, and particularly in the social animals living in the ocean, a common strategy for combating predation is sociality because it is one way of securing safety in numbers [94]. Many researchers are arguing that social and collective behavior started in the ocean because there was nowhere else where to hide and the ocean provides the ideal environment and ecosystem for cooperative and social behavior.

Most research on trust and reputation mechanisms in society is inherently linked to studies of cooperation and reciprocity and trust and reputation are another mechanism to secure safety in large interactions. In general, both economists and sociologists agree that reputation networks are stable and take a very long time to form. Particularly with respect to informal value exchange networks, trust and reputation mechanisms function as intrinsic forces in a community and society for monitoring "bad" behavior such as cheating, deceiving, free riding, and many more. In a way, trust and reputation mechanisms have evolved in a network or community as a way to stabilize and sustain these networks.

An example of such collective communication that involved the trust of the community was the nilometer. The "nilometer" is a device part of the temple complex in Ancient Egypt. What the nilometer did was to compute taxes—the

A. Berea, *Emergence of Communication in Socio-Biological Networks*,
Computational Social Sciences, https://doi.org/10.1007/978-3-319-64565-0_3

higher the harvests, the higher the taxes. In many ways, the ancient devices used for communication were linked to economics—writing in Mesopotamia emerged due to solving accounting recording problems, while in Ancient Egypt the nilometer was a signal regarding the fertility and welfare of the land. Another such example is that during the Ptolemaic era, the Ptolemaics who were of Greek descent, would carve their names in the well as a measure of accounting for their contributions to building up a well.

Currently there are debates about where the first recording of information as writing happened. While some researchers believe that the information recording as writing first arose in the Mesopotamian area and era, with the cuneiform writing of the tablets from Uruk, this idea has been challenged by an even older form of writing discovered in Eastern Europe, in what is called by historians the Tartaria culture. Although these objects have been discovered at the middle of the twentieth century, only very recently the meaning of their symbols has started to be understood [55, 66]. The current hypothesis is that the first writing was inspired by religious practices with cosmological symbols and not by economic necessity, as the clay cuneiform tablets did and that these first information recording objects were used to depict communication as a thought process and not only as natural, evolutionary signaling.

Some of these early accounts in the history of our civilization regarding collective economic behavior show how intrinsic to the development of writing and the means of recording and exchanging information the collective human endeavor has been. But the way we form collectives or scale up in a non-linear way, when the scaled-up system morphs into a new entity, is not very well understood. This is also one of the major features of complex systems, called near-decomposability [80]. Near-decomposability represents that property of a system similar to divisions and partitions, yet different in the way the parts of the system function and communicate together, i.e. the organs in the human body versus the parts of a mechanical engine. Near-decomposability makes the parts of the system as well as the parts and the whole system communicate in a very different way from how the parts of a system communicate in a fully designed, decomposable system. In other ways, we can say that near-decomposability is a property where the most interesting communication related to scalability and collective behavior happens.

The "intelligent" collective behavior, whether organized or spontaneous (ad hoc groups, cities, firms, crowds, countries) has been coevolving with communication and our language for a long time, yet we don't know enough about how the evolution of collective behavior in humans changed human language and vice versa; we know it has changed technology and the development of better means of communication, we recognize that it has also changed the vocabulary and dictionary (the English language vocabulary has increased more than 10 times in the past few years), but there is no clear understanding about how these coevolved together.

3.2 Patterns of Communication

The goal of this research is to explore the effect of different informational structures (networks vs. symbols vs. images vs. text vs. probabilistic vs. Boolean vs. sound vs. any other form of information representation) into the evolution and emergence of communication patterns. These patterns are necessary to create a taxonomy of communication that includes both the organic, natural and designed, artifactual communication (and any mix in between) as a better representation of the physical and informational worlds.

We often confuse the meaning of communication (the content of the message) with the means of communication (the way the message content is being transmitted—chemical, visual, digital, a.s.o.) and this confusion is reminiscent from the way we theorized information exchange for the purposes of analyzing and quantifying it, in bits and without any meaning [78].

Few people are aware of the fact that the textile industry and the computer history have a lot in common. Particularly, the repetition of patterns that is required in the textile production of scale was at the basis of Babbage's punch cards and the very first computer algorithms. The repetitive nature of patterns that we find in fractals, in visual arts, in nature and in computation is essentially a method through which we can record and transmit information. Jacquard, when he invented a new way of waiving textile patterns, he basically invented the punched card. Information was translated from pictures into the final products. This revealed the power of abstracting information and creating representations of information that could be scaled up.

Another great example of how textiles industry and information theory merge together in patterns is also the Brocade in Lyon. In this great work of art, tapestry becomes a form of storage of information on textile fabrics.

Statistical regularities such as the Zipf distribution (observed not only in languages, but also in the distribution of firms or human organizations) or the edge of chaos or cellular automata rules or the entropy patterns are great examples or a compass for what constitutes intelligent language and intelligent patterns of communication. But so are the more trivial symbols and signals, such as company logos, financial trading time series, Internet memes, and accounting bookkeeping. Accounting bookkeeping is distinctly interlinked with the emergence of writing, the patterns in the textile fabrics are linked with the emergence of computer punch cards and the emergence of telegraph with the expansion of the railroads. The bottom line is that the emergence of artifactual means of communication in human civilization is closely intertwined with the emergence of entreprenurial and economic systems of our civilization.

The human species created these "things" called organizations (from clans and cities and firms and political parties and financial markets and many more) that represent, in one way or another, experiments in designing collective human behavior. They are an inherent part of what constitutes our civilization and they would not have been created or sustained without communication (Fig. 3.1).

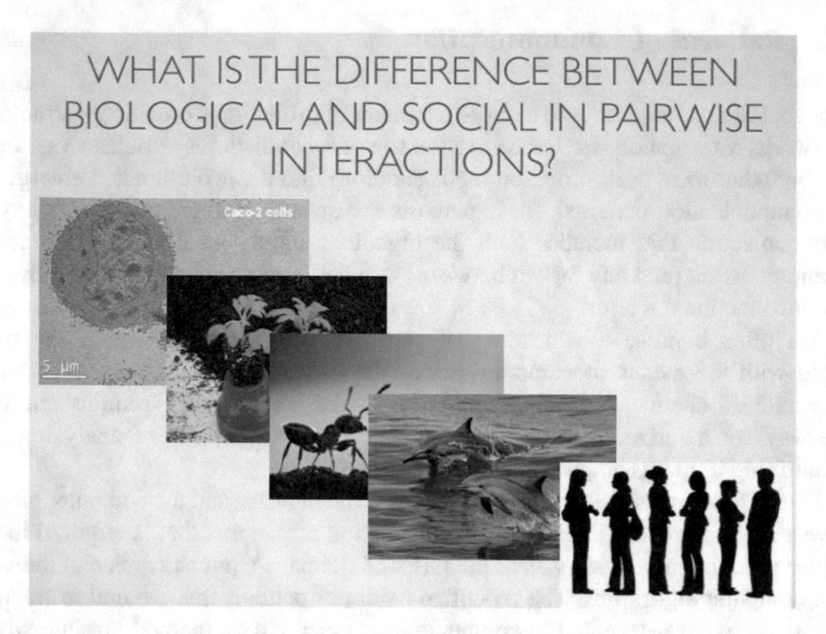

Fig. 3.1 Pairwise interactions in information exchange and communication. The scalability of communication and the exchange of communication across living species depends on pairwise interactions

We know that language has largely evolved organically and endogenously and with few exceptions (the nineteenth to twentieth century rules of linguistics imposed by the academies in a few European countries), language is essentially an informational representation of the society and culture where it was used. Before digital means of communication were invented, the economics of the world was localized, embedded in the communities and the culture of the societies (salt in North Africa, silk in China, wine in France, tourism in Italy, a.s.o.) and with large variations in geography. On another hand, we have designed informational systems that are universal to some degree (mathematics, computer languages), but devoid of any meaning or content, and therefore used only in small human-computer, high-tech communities. One organic language has become a standout among others due to physical trading practices (English) and first protoglobalization phase in our civilization, while in the digital, computer based world, various computer languages are still competing for global hegemony and this effect is largely due to the fact that computers cannot choose the language they are operating with by themselves (yet). Therefore there is a lag between the human use of computer languages and the human use of human languages and a lag in the physical and informational economy that has the potential to create societal crises. In an economy that will continue to be dual physical-informational (at least until we discover how to trade information for energy and matter and vice versa), the role of a communication system that blends organically with the humans, the environment, and the digital world is probably going to be a game changer in global innovation.

To what degree can the current designed communication evolve organically (bots inventing their own language) or become organic and to what degree does the organic communication become standardized (English words imports into other languages, Internet memes, etc.)? A map or a clear picture of the communication patterns as they were in the past and are today would help us understand the evolution of communication in the near future (will there be a new type of human-computer language invented that would change the paradigm of the current software languages that have standardized the way humans interact digitally?).

Language (in the broadest sense) has deep effects into the selection and adaptation of a species, whether this species is ants, dolphins, or humans, and the more social a species is, the more important is its language for survival. But in the case of a civilization where the biological evolution standards of selection and adaptation are more diffuse (we don't really know why we create art or why we are creating technology that make us more comfortable, more individual and less fit for survival), while the social evolution standards are more acute (we compete not for resources, but for networks of influence and power, whether these networks are firms, associations or friends), communication patterns that give more fitness to networks of influence would be the ones that take advantage of both the organic and the designed patterns of communication.

What is the difference between models and language? This is a very interesting question that has not been asked very often. Both models and language are representations of reality—models are "designed" representations of reality, since the modeler (usually the scientist) needs to put some thought into what features or what phenomena he or she wants to extract from reality and study, while language is an organic, endogenous, and social representation of reality—it is a system of norms and reality representations that has developed throughout a very long period of time and through multiple interactions and information exchange of all the people. Scientific models also have a tendency to evolve and build from each other, discarding others or helping to build others—but scientific models are relatively new and are accessible to very few people in specific areas of study. Scientific models tend to be more rigorous and they do sometimes lead to "scientific jargon" or specific expressions (i.e., "asymmetric information" certainly did not exist before economics was invented). Not all scientific models are mathematically based, but most of them are, while all of them follow certain ruled of design. On another hand, languages, while they also evolve and borrow new words from other languages and discard older, no longer used words, are not designed, do not follow specific rules of representing reality, and are inherently organic. Either scientific models or languages take a long time to learn and evolve, but for scientific models we need to use our rationality and critical thinking, while for languages we need to use our perceptions and intuitive thinking. Therefore scientific and language patterns are expressions of our inner and outer information.

Mathematically, the most beautiful expression of the idea of two dimensional patterns that are used in communication is represented by cellular automata models. Cellular automata models show emergent patterns out of very simple rules of behavior, visually similar to the textile jacquards, yet they are emergent and become new entities based only on their own contained rules of communication (see Fig. 3.2).

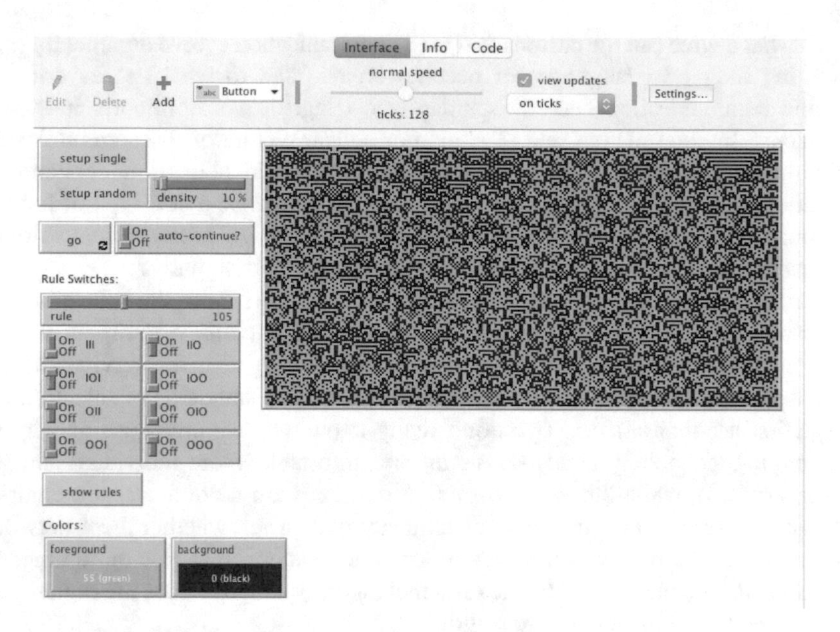

Fig. 3.2 An example of a cellular automaton from NetLogo models library

3.3 Beyond Information Diffusion Models

The problem of information exchange in a social network has been studied by computer scientists, econophysicists, epidemiologists, social scientists, and especially by economists, such as Hayek (as mentioned above), but also Stiglitz, Hirshleifer, Akerlof, and many more [1, 41, 86]. Older models of information diffusion or informational cascades in a network analyze information primarily with regard to the strength of its ties (weak, strong, or normalized). In a highly influential paper, Granovetter showed that the weak ties in a social network are more relevant for information diffusion than the strong ties [33].

Another influential paper is Friedkin's [30] study on informational flows. He showed that in academic networks, the information flow is highly dependent on the weak ties that the organization has with the outside networks.

On the other hand, Carley's work on information diffusion and network topology takes into account the possibility of asymmetric relationships and "cognitive inconsistencies" [50]. Also, basic socio-cognitive mechanisms, when carried out across and within large numbers of individuals, result in the coevolution of social structure and culture [18]. She also refers to an ecology of networks, which is a collection of knowledge networks between agents, formed by interactions and joint decisions. These interactions have also been studied broadly by Hannan and Freeman [38], within their work on organizational ecology. They argued that in a social network, selection is more important than adaptation and that there is a high dependency on the network density and resource partitioning.

In Axelrod's [2] cultural model, there is informational heterogeneity introduced through the five features of culture, and the traits of each feature. Also, in his model, the successful dissemination of culture presupposes a degree of similarity with one's neighbors; it is spatially bounded and essentially a cellular automaton type of information diffusion. One interesting example of information exchange is Knocke's model of people and events [49]. In this study, information is homogeneous and refers to a static collection of events. The study looked at the embedding of the organizations in the network based on how close or apart they are regarding their social welfare views.

Extensive research on the diffusion of innovation in social networks has shown the importance of social contacts in determining behavior [20, 71]. Rogers has shown how the spread of Internet has changed the way people communicate and adopt new ideas. Some basic insights from the models of diffusion are: the conditions for lowering the thresholds of diffusion [90], the probability of transmission on any given link [46], the role of degree density in transmission rates, clustering and degree correlation [60, 61]. The most important aspect that is neglected in these researches is that networks often actively react to the ongoing process [45]. Also, if information is valuable, individuals start searching through the network specifically or create new links and the problem of how agents and networks adjust to diffusion is still open [45].

Economic theories of information focus on the peculiarities of information as an economic good and on asymmetric information problems in a market setting, but overall they are treating information as a homogeneous good. Both heterogeneity and asymmetric information raise important economic problems (see Fig. 3.3).

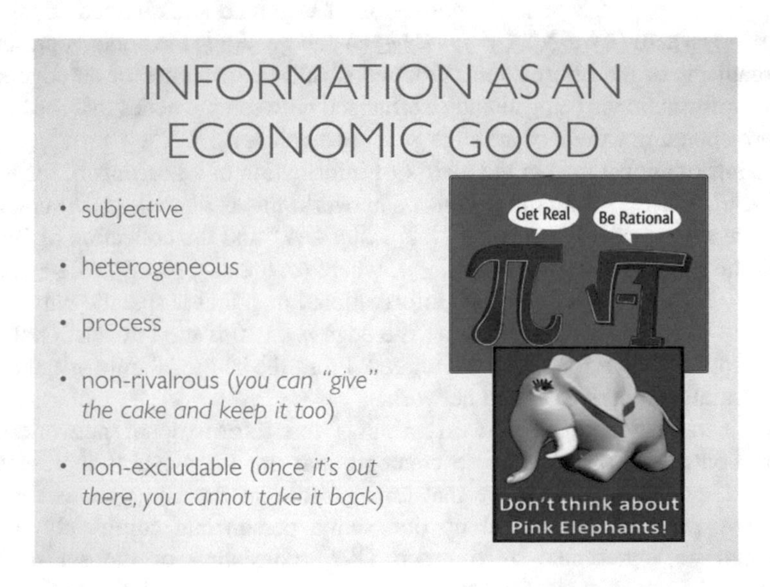

Fig. 3.3 Information as an economic good. The properties of information as an economic good are subjectivity, heterogeneity, processual (not entity), non-rivalry and non-excludability

There is actually very little economic literature with respect to heterogeneous information. Information is rather asymmetric, as heterogeneity poses too many problems with respect to future expectations, forecasting and Bayesian analysis [1, 42, 85].

Information can also be "sticky" [56], with the property of slow dissemination in the labor market. Information frictions in the macroeconomic literature are basically specifications for building models of slow information diffusion in the economy, with consequences for unemployment and economic growth.

On another hand, Searle's ontological subjectivity is an important concept that makes subjective value of heterogeneous information understandable and computable in a social network setting that is not necessarily a market [75].

Dodds et al. [23] have proposed a model of information exchange in organizational networks, with efficiency and robustness properties for adding ties. Their analysis on ultra-robustness and endogenous/exogenous causes showed that organizations with non-hierarchical ties are more likely to survive disasters.

3.4 An Alternative to Shannon's Model

A theoretical framework for modeling heterogeneous information exchange. In a bimodal network of persons and information, the networks of people and personal information maps are represented as follows:

(a) the organisms network as a graph (P, g) consisting of the set of nodes $P = p_1, p_2, \ldots, p_n | n \in N^*$ and the collection of weighted and directed edges $g_{ij} = (p_i, p_j)$, where $(i, j \in N^*; i \neq j; \forall i, j \leq n)$ and $g_{ij} \neq g_{ji}$; the nodes represent the organisms or the information "producers" and the presence of an edge shows that information is being already exchanged between the nodes that are tied; the corresponding weight of the edge g_{ij} takes a value, $w_{ij} \in N^*$.

(b) the informational map or the matrix of information of an organism: for $\forall$ node $p_i \in P, (i \in N^*)$ from the organisms network, this is a graph (Y_i, h) consisting of the set of nodes $Y_i = y_{i1}, y_{i2}, \ldots, y_{im} | m \in N^*$ and the collection of Boolean and undirected edges $h_{ab} = (y_{ia}, y_{ib})$, where $(a, b \in N^*; a \neq b; \forall a, b \leq m)$ and $h_{ab} = h_{ba}$; this is the subjective informational map that is specific only to node p_i and the corresponding value of the edge h_{ab} is 0 if the informational nodes are not related and 1 if they are related. From the bi-modal network, there can potentially be derived a third network:

(c) the meta-informational network: this is the informational network of all the nodes that connects those concepts that are identical. It represents the social, collective knowledge that emerges out of the endogenous exchange at any given time and that no one single person can completely possess. It can be represented by a graph (Y, h) consisting of the set of nodes $Y = \{y_{11}, y_{12}, \ldots, y_{1m}, y_{21}, y_{22}, \ldots, y_{nm} | n, m \in N^*\}$ and the collection of Boolean and undirected edges $h_{ab} = (y_{ia}, y_{ib})$, where $(a, b, i \in N^*; a \neq b; \forall a, b \leq m; \forall i \leq n)$ and $h_{ab} = h_{ba}$ (Fig. 3.4).

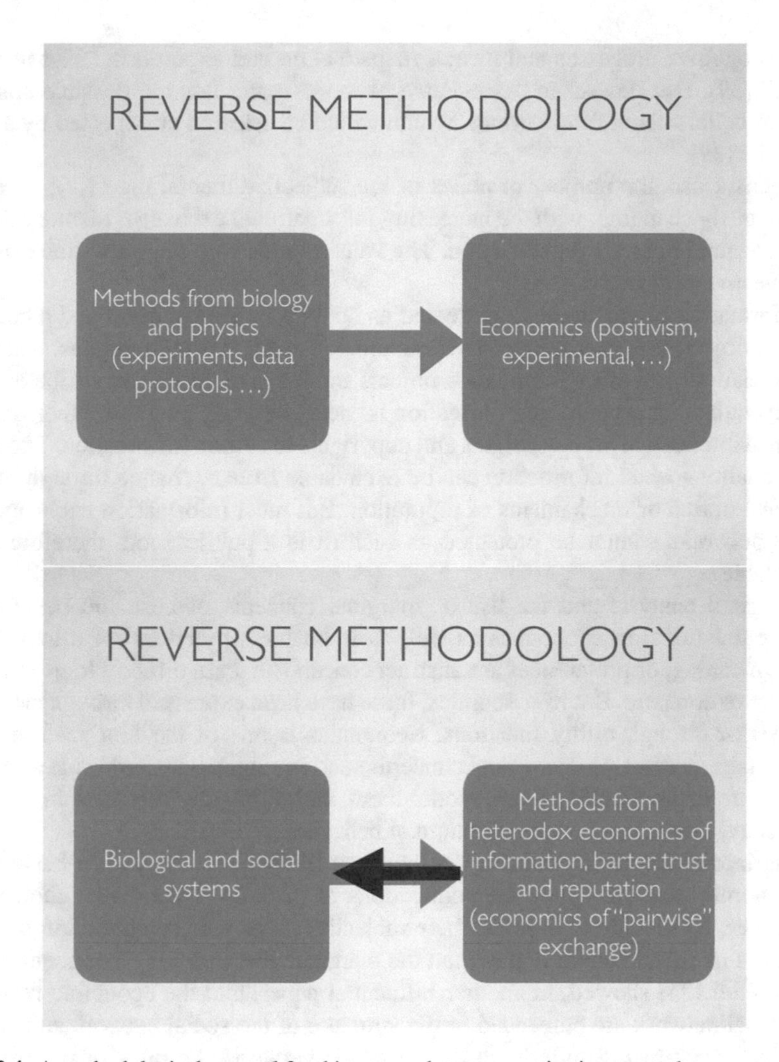

Fig. 3.4 A methodological reversal for this approach to communication research

The nodes p_i of the social network are also the acting agents in the model of exchange. The information is the heterogeneous object of a person and it cannot exist outside the social topology. The heterogeneity of a person is implicitly given by the heterogeneity of the subjective informational map. Any person p_i has three functions: produces (internally), exchanges (through signaling or search), and loses information (through memory decay). These functions are subsequently detailed.

The informational map of a person or the mental map is a subjective informational set that changes continuously with incremental learning and memory decay. It is represented as a symmetric Boolean network that roughly approximates the association processes in the brain. This is in accordance with Hayek's interpretation of the cognitive processes and mental maps that he had exposed in “The Sensory Order” [39]. For Hayek, an “associative process” represents the possible connections of beliefs about future events, which would be selected or expected by a map or model [39].

In this sense, the number of nodes of the subjective mental map ($|Y_i| = m$) is continuously changing, with the increasing informational exchange, memory decay, and personal information production. The value of m is very large and finite from a p_i subjective perspective.

Information in this network is treated as both an economic good and a computational object: it is near non-excludable and non-rival, it is autonomous, has rules of association with other information objects and it has methods and attributes. The non-excludability property of information is “near” because sometimes there can be assigned intellectual property rights and copyrights to certain information. There are also situations when information can be excludable from exchange through means of social norms or mechanisms of reputation. But most information exchanged in social networks cannot be protected as such (it is a public good, therefore non-excludable).

Marginal analysis and the use of marginal concepts of cost and benefit has been foundational in economics as well as in business practices for a long time. Value, meaning, or preferences are abstract concepts that are difficult to quantify or measure or compare. But in economics, these have been expressed through marginal analysis or through utility functions. Economics is one of the first sciences that tried to take abstract, philosophical concepts such as subjectivity and value and tried to measure them, quantify them, model them, and ultimately use them in order to understand fundamental aspects of human behavior.

The informational relations or ties between the persons follow the basic rules of economic exchange: they are connections of persons at their own choice and preference, by seeking to improve their subjective maps with semantic information that has a marginal benefit higher than the marginal cost ($\delta B > \delta C$) of acquiring it. Granovetter [34] showed, in another influential paper, that the economic relations of a social network are embedded in the structure of the social network and do not exist in the abstract. Similarly, in this semantic network, the rules of information exchange are economic.

Time is modeled discretely (1 time unit = 1 series of simultaneous exchange). Also discrete is the economic marginal analysis (the marginal benefit and marginal cost represent discrete changes—$\delta B - \delta C$—and not differential equations). This is methodologically important for the computational model, where the economic good “information” is an object and the marginal benefit and marginal cost are its attributes.

The heterogeneous economic characteristic of information is to have heterogeneous marginal benefits and heterogeneous marginal costs for each informational set that a person exchanges at one time. Computationally, heterogeneous information refers to subsets that overlap across individuals [11].

These informational subsets are graphs ($Z_i, h|_z$) consisting of subsets of nodes $Z_i \subset Y_i$ and the network $h|_z$) restricted to the subset Z_i, so that $[h|_z)]_{ab} = \{(1 \text{ if } a \in Z_i, b \in Z_i, h_{ab} = 1)|(0 \text{ if otherwise})\}$. This is the network that represents the information exchanged from p_i to another node at one time, and thus it is obtained by deleting all the links except those between the nodes of the subset Z_i. Some elements of the subset Z_i may be identical to some of the informational map $Y_j(\forall j \neq i)$ of the receiving node p_j ("shared" concepts of the two nodes). The receiving node p_j alters his informational map accordingly, by creating new relations of association between the existing nodes in Y_j and the non-identical nodes acquired in Z_i.

One of the challenges with interpreting and visualizing these "nested" networks comes from the nature of the edges; they are information as well. The connectedness of any two given nodes, whether it is a personal, a business or other type of tie, means that some sort of informational exchange between those two nodes has already taken place. Any edge/connection has an implicit information inventory.

Due to the near non-excludability property of information as an economic good, the exchange of information is a directed tie from the person that receives the information to the person that produced the information (receiver $\leftarrow$ sender). Whether node pi receives passively information from p_j (a signal) or searches for information from p_j, the informational subset $(Z_i, h|_z)$ that p_i acquires has a specific δB and δC. Therefore the values of δB and δC are assigned by the receiver with each informational subset it acquires.

For the semantic analysis of this paper, information is distinguished into two types: noise and "meaningful" or valued or semantic information.

Then $\delta B = 0$ for noise and $\delta B > 0$ for meaningful, semantic information. The more important the information exchanged is for node p_i, the higher the δB. The cost of acquiring any piece of information, if it is being searched for, is $\delta C > 0$, while if it is a received signal, is $\delta C = 0$. The same piece of information may be meaningful for one person and noise for another.

Due to the non-rivalry property of information as an economic good, the sender does not give up that specific informational subset. Thus, the subjective mental map is modified only by:

(a) the production of new information
(b) the exchange of information through signaling or search
(c) the memory decay function (see Fig. 3.5).

The production of new information is left for the moment as a "black box" or as an unspecified function. Within this black box it is also left the information that comes from the interaction with the physical environment (such as perception or observation). From an economic perspective, the difficulty of assigning computational values to the informational production function comes also from the difficulty of assigning property rights to it. If it is hard to measure the subjective incentives for producing new information, then it is hard to formulate it mathematically.

This does not impede the model of exchange. For the purpose of informational exchange, the computed subjective δB and δC incorporate all the necessary information with respect to the produced information.

Memory decay is described by the psychology and neuroscience literature as either a power law [95] in [97] or a logarithmic function [97, 98]. But the common agreement is that the rate of memory decay decreases with time, meaning that more

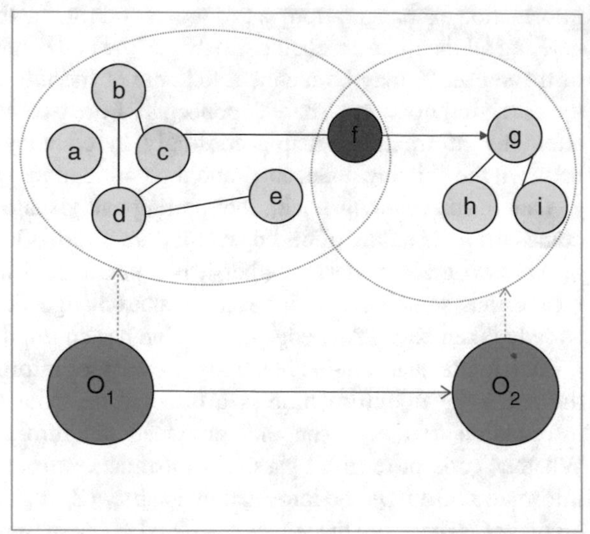

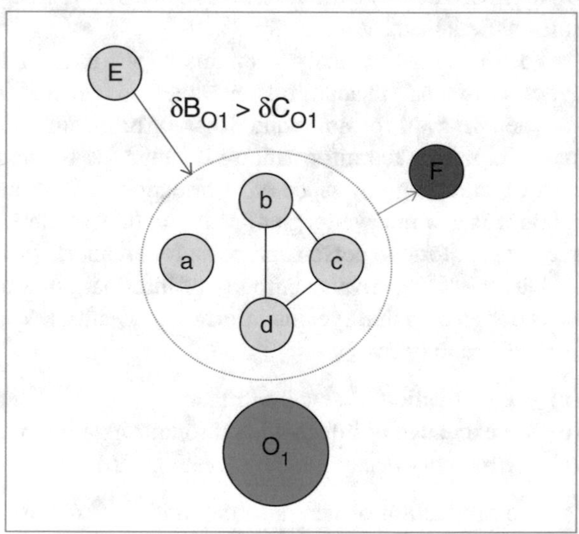

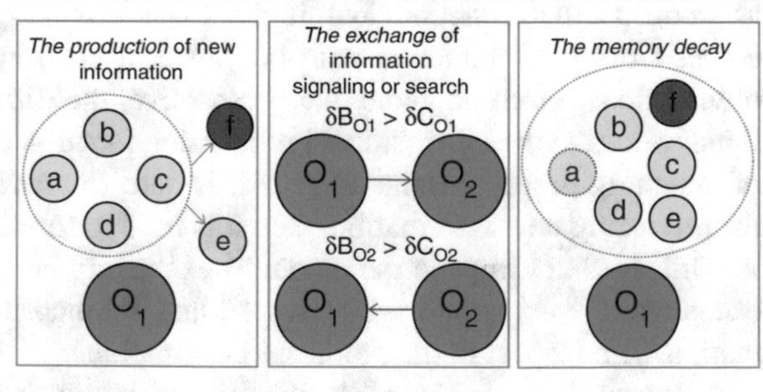

Fig. 3.5 The economic model of communication from pairwise interactions to a scalable process

consolidated information is less forgettable. In terms of this network, the rate of decay is higher for the information that is closer to noise ($\delta B = 0$) and smaller for the semantic information that is very meaningful ($\delta B >> 0$).

The social network is endogenous and all information is produced and diffused endogenously. An endogenous network is an environment where the interaction structure is endogenous. The endogeneity of the network in this model is given by two rules: the endogenous exchange (edges) and the endogenous preferences for information (nodes "choose" their values for δB and δC). Information does not exist in the abstract, outside the social network and it has no meaning for our network without the human interpretation of this information [75].

People usually produce and seek to acquire "meaningful," semantic information and only accidentally produce and consume "noise" (it does not pay-off to do so, $\delta B = 0$) (Fig. 3.5).

3.4.1 Model Description and Behavior

The model is a representation of the previously described exchange and the evolution of the bimodal network over time. It can be instantiated with a finite number of agents, networked in a specific topology. The three types of topologies that will be tested for semantic emergence are: Erdos-Renyi [27], scale-free [3], and small worlds [93].

The agents in the time evolving model are the nodes p_i of the social network, (P, g). Each agent p_i has as attributes an array of δB_i ($\delta B_i \in [0, \max(\delta B)]; \forall i \in N^*, i \leq n$), an array of δC_i ($\delta C_i \in [0, \max(\delta C)]; \forall i \in N^*, i \leq n$) and a rate of memory decay, d_i ($d_i \in [0, 1]; \forall i \in N^*, i \leq n$). Since in this model the memory of a node is not represented numerically (the absence of a production function), the rate of decay will transfer to the exchange process: meaningful information has a rate of decay decreasing from the most meaningful and noise has the rate of decay $d_i = 1$.

The agent chooses the value of δB_i after inspecting the number of nodes in Z_i that are identical to the ones in its own informational set, Y_i: the higher the number of identical nodes, the higher the value of δB_i. He also chooses the value of δC_i after inspecting the number of nodes in Z_i that are non-identical to the ones in Y_i: the higher the number of non-identical nodes, the higher the value of δB_i.

There are two types of directed interactions/exchange in the network: (1) signaling; (2) search for information. The agents choose randomly to either signal or search for information at each time step t.

Given the informational map (Y_i, h) of the node p_i, when the node sends information (a signal), this signal is a $(Z_i(t), h|_{\llcorner}(t))$ subset at time t ($t \in N^*$). The sender p_i chooses a cost $\delta C_i > 0$ of signaling to p_j. He also chooses an $\delta B_i > \delta C_i$ (incentivized signaling). For the receiving node p_j, the cost of receiving this signal is zero ($\delta C_j = 0$) and the benefit δB_j can take any value between 0 (when it is noise) and the maximum value of δB_j (very meaningful information). The agent p_j chooses the value of δB_j from its own array.

When the receiver p_j searches for information from the node p_i, it invests a cost $C_j > 0$ and receives a benefit $B_j > 0$ (it does not search for noise). The general condition for incentivized search is that $B_j > C_j$. The information exchanged in this case is another $(Z_i(t), h|_z(t))$ subset at time t. For the sender p_i in this situation, the cost is zero ($C_i = 0$) and the benefit B_i can take any value between 0 (the node does not benefit from giving away the information) and the maximum value of B_i (the sender can also benefit from giving the specific information, as in the case of signaling). The agent p_i chooses the value of B_i from its own array.

As initially defined, the corresponding values of the edges g_{ij} at time t are $w_{ij}(t)$.

Having a certain initial topology of the social network (and not "growing" the network) implies that the connected persons have already exchanged information, while those that are not connected do not know about the existence of each other. The values of the initial edge weights $w_{ij}(0)$ (for $t = 0$) are assigned as following a fitted distribution.

The inventory of exchange under this theoretical framework is represented by a cumulative function of the repeated informational exchange between any two nodes. Edges are information as well and the weight of an edge after 1 time step is defined by:

$$w_{ij}(0) + (1 - d_i(1)) * (\delta B_i(1) - \delta C_i(1)) + (1 - d_j(1)) * (\delta B_j(1) - \delta C_j(1)) \\ = w_{ij}(1), i, j \in N^* \tag{3.1}$$

where all measurements are information. Or, in general form:

$$w_{ij}(t-1) + (1 - d_i(t)) * (\delta B_i(t) - \delta C_i(t)) + (1 - d_j(t)) * (\delta B_j(t) - \delta C_j(t)) \\ = w_{ij}(t), i, j, t \in N^* \tag{3.2}$$

After a number of time steps t, we look at the values $w_{ij}(t)$ for each edge. We also extract the values δB exchanged and compare them relatively to $w_{ij}(t)$:

$$\rho(t) = \left[\sum^{t}\sum^{i} \delta B_i(t) + \sum^{t}\sum^{i} \delta B_j(t)\right] \Bigg/ \left[\sum^{i}\sum^{j} w_{ij}(t)\right], \ \forall i, j, t \in N^* \tag{3.3}$$

This comparison (the value of ρ) will show how much of the information exchanged has a high semantic dimension and how much is noise over the entire social network. Performing these comparisons for each of the three topologies as well as for different time steps, t, we can determine if and perhaps how the topology matters for semantic emergence. Higher values of ρ show that more semantic information has been exchanged in this network. Also, comparing the values of the ratio at different time steps t will show if the model is sensitive to time.

This theoretical framework and model description represent a proposal for studying heterogeneous informational flows or information exchange as an alternative to information models that are quantified by the amount of information, instead of the value of information. This is obviously a very crude, toy representation of how such exchanges or "trades" of information happen. But with the power of computational models, we can step further into understanding whether patterns of trade or markets for information can occur given that each of the living organism has a unique experience and way of interacting with the world.

Current literature avoids the analysis of complex problems such as heterogeneous information and semantic interpretation of information diffusion due to the difficulty their nature poses to standard methods.

The theory and model outlined in this paper have the purpose to look for semantic emergence. This is not a model of information diffusion. Usually, information is analyzed in relation to time and uncertainty [42]. While this theoretical model leaves uncertainty aside, better insights surely can be drawn from analyzing information exchange by adding heterogeneity and time.

Perhaps the most important plus that can be added is the cognitive modeling of the agents as producers of information and learning agents. This will no doubt provide significant insights into the emergence of semantic networks.

Chapter 4
Constructed Language Versus Bio-chemical Communication: An Agent-Based Model and Applications

4.1 Communication and Morphogenesis: Cellular Beginnings

The movement of information is the basis of biology. Life happens and creatures evolve because information is transferred [94].

If we are truly looking at a fundamental research of communication, we need to address an epistemological problem: how do we connect the subjective phenomena with the objective existence? All phenomena arise initially from sensory information or from memory of such information; in the cell, this includes information from the environment plus genetic information. Variation in genetic information is ex-ante, variation in cognitive information is ex-post. We know from cell morphogenesis research that there are two types of memory: genetic/physical and cognitive/symbolic [65].

Rene Thom, who was the proponent of catastrophe theory, also looked into more depth into problems of morphogenesis and origins of language, particularly into parallels of biological morphogenesis and semantics [88]. In the same tradition as Turing's seminal work on morphogenesis, Thom attempted to show that morphogenesis can be described mathematically and that languages can be described mathematically and morphogenetically as well. Thom also relies on previous work on morphogenesis that had been done by Waddington on creodes and stable processes in genes and cells [92].

To understand Thom's theory better, we need to think morphogenetic processes and morphogenetic fields in terms of stability and stable processes. In this way, morphogenesis "works" somewhere on an edge of "chaos" and "order," to put it in Stuart Kauffman terms [48]. Thom is also advancing the idea that a morphological process is a semantic model and "a kind of generalized m-dimensional language," and language is a subset of this, with just one dimension [88].

A. Berea, *Emergence of Communication in Socio-Biological Networks*,
Computational Social Sciences, https://doi.org/10.1007/978-3-319-64565-0_4

4.2 Adaptation and Selection in Communication

Another important question to ask ourselves is what is the difference between language and reality and between language and subjectivity?

These questions are not asked solely for the purpose of an exercise in abstract thinking. We use language in order to describe and understand our own reality and we know that language and context are intricately linked together. Similarly, we use language in our own contexts and biases and the use of languages reinforces or dismantles some of these biases and therefore our use of language tends to be highly subjective as well. For example, an engineer is most likely going to shape his or her world view through the personal experiences of working with exactness and usefulness, and interpret the language through this lens. Similarly, someone born in Germany is more likely to shape his or her worldview and biases through the German culture. This is true for the human species, but to some extent it is true for any other living being on Earth—if this being is communicating with chemical signals, then the chemical reality around its physical space is going to be influential to its communication and its "language" or chemical signals, on their turn, are going to be shaped according to how it has adapted and survived its immediate physical context.

In human languages, the most stark evidence of language selection given the reality and subjectivity contexts is given by the disappearances of languages simultaneously with the disappearances of cultures. Currently there are some efforts to digitize and record the nearly extinct languages for future generations. The fact that the number of languages in the world is shrinking seems to be an irreversible trend, in line with our more globalized, more integrated world. As with any other aspects of our lives that go extinct, particularly the plant and animal species that go extinct, so do human cultures and human languages. There is even a dedicated word—"linguicide"—that depicts the death of a language that occurs once the last native speaker has died.

The language loss has been correlated with economic growth, perhaps due to the fact that economic growth is also correlated with globalization and the increase in urban population and cities.

The fact that languages evolve is as inexorable as the evolution of our society. The efforts to preserve languages, and cultures, can help record the evolution of our civilization, of our history of subjectivity and meaning and hopefully also give us more insights into the future of our communication as an intelligent, cultural species that constantly evolves and adapts economically, socially, and biologically.

The concepts of adaptation and selection from evolution and complex systems are somewhat similar to the concept of efficiency and competition in economics.

The fact that human language is an emergent phenomenon is quite largely agreed upon by the linguists and the other scientists researching language. But human language is not only an emergent phenomenon, arisen from endogenous rules and many years of practice. While this has been true for most of human history and civilization, in the past couple of centuries we had experts and scholars and academias in the Western world that would impose rules of usage of the language and rules of writing. The dictionaries were born from the need to standardize and design parts of the language for proper usage.

This did not happen for all the languages—most of the human languages did remain embedded in their cultures and the smaller or more isolated the culture, the less likely it was to be influenced by experts and the imposed rules of usage and writing. In many ways, the imposed rules of language by the experts were more than guidance and these spurred new questions about the evolution of language, such as: How much design and how much evolution? How much preservation in language and why? Should we preserve our language as it is or just let it evolve? Can we preserve it at all? The evidence of the rate of dying languages versus the preservation efforts shows that we are certainly losing more than we will ever be able to preserve in human history, and thus there are languages that will never be recovered or ever used again.

Similarly in animal species, cultural transmission can cause variation in behavior within a species. It is not enough though simply to demonstrate that animals in one place, or one group, behave differently to those in other places or groups. Environmental variation can also produce differences in behavior between communities in several ways. For instance, if animals can learn a behavior with no social input, then in places with a particular food or tool present these may get used, whereas their use will be absent from communities where the necessary food or tool does not occur. This behavioral variation is not culture; it results from individual learning in different ways on their own. Genetic variation can also lead to differences in behavior [94].

But what is the difference in selection and adaptation between human languages and animal "languages"? Do animals have a very efficient language or a language that is more efficient in the exchange of information than humans do? Do humans have more noise and therefore more quantity of communication in their language?

One situation when we had to communicate in the most efficient and most meaningful way has been when we had to transmit "the essence" of our culture and language into the Universe, to an unknown destination and unknown recipient, through the Voyager and Pioneer plaques. When Carl Sagan had the idea of transmitting the most essential features of our culture and civilization through these space probes, humanity had to come up with the most universal, general but also most meaningful messages, encoded as words, mathematics, sound and images—basically in all means of information transmission we currently have. These messages still contain some of the most essential means and meanings in our civilization and therefore very efficient, condensed, and meaningful information of our humanity. The big question is: can someone else decode it, understand it, and react to it or is it doomed to remain purely subjective to our human context?

At a smaller scale, we know that ideas spread faster between people that understand each other's biases (that share similar cognitive bias). If such bias-in other words, subjective context-is not shared between the sender and the receiver, then is the same "meaning" of information being exchanged or is the information exchanged actually conveyed with different meanings? And how much of the meaning versus the means of information is actually used for adaptation and selection of communication by either the sender or the receiver? Is it possible that the mode of storage and transmission of information is actually even more important than the meaning of information for selection and adaptation processes? Is the mode of transmission and storage the one that actually determines information meaningfulness and interpretation due to adaptation and selection mechanisms of the receiver?

Lera Boroditzki showed that language does change behavior [10] and Steven Pinker showed that grammar (language structure) and language evolution are shaped by the transmission protocols and conventions of communication that are also shared by the sender and the receiver [67]. In other words, we need to pay attention not only to the adaptation and selection of the words and messages, but also to the adaptation and selection of the means of communication and to make the problem even more complex, the two are inexorably linked together.

The way we think about this problem has deep consequences into the way we think about social behavior in organisms and economic behavior in humans. Communication does not only help us act, but the way we use communication shapes our beliefs and therefore our actions; which came first, communication or action? Our decisions informed by our language or our actions are informing our language and our decisions as well? The answer is, of course, both.

We observe this coevolution of the means and meaning of information exchange not only in human languages, but also in the abstract languages and computer languages as well—the usage and adoption of a language is reshaping the language syntax and coding norms. And at the biological, fundamental level, memory that is encoded in more senses will live longer ("living" information than information that is simply stored in one type of information recording)—for example, memories that we encode visually and olfactory will remain with us for a very long time.

4.3 A Basic Agent-Based Model of Communication

Communication is one of those versatile processes that can be as heterogeneous as the agents and as adaptive and evolving as they are. One way we explored some of these ideas outlined above at the University of Maryland is through a mathematical (theoretical) model that includes the properties of heterogeneity and subjectivity of information, described in the previous chapter and then through a computational model (simulation) where we could track different fundamental scenarios of communication. We created a model of communication (information exchange) that can be adapted to cells interaction, organism (animal) signaling, humans languages, human and computer interactions and we showed the differences between organisms that have the ability to change the rules by which they exchange information (decision) and organisms that only react to information stimuli.

Some of the earliest research on semantics and meaningful communication started in the early stages of the AI research. One such example is the work of Ronald Brachman, who developed at Yahoo the early models of semantics in networks and even a computer language for including semantics in the exchange of information, called KL-ONE [12]. This was one of the oldest ontological languages (knowledge representations) and it included deductive reasoning in order to discover new information about the ontology.

Before natural language processing and before current machine learning techniques that are processing texts and language, a precursor for modeling language and communication has been the niche field of mathematical linguistics. Mathematical linguistics has been more concerned with modeling human language in a structured form in such a way that any sentence could essentially be "translated" into math and performed mathematical operations upon it [57]. In other words, mathematical linguistics is attempting to discover a meta-language of languages or a universal language. This is quite different from the information theory model from Shannon, where information, not linguistic syntax, is being quantified.

Our model presented here is abstract and general enough to assume from the mathematical and computational viewpoints a similarity between the connections in the brain and the exchange in the market; prices are only a subset of information and trade is only a subset of pairwise connections.

4.3.1 What is the Agent-Based Model?

This model is an implementation of the research design described by the previous chapter and seeks to model information exchange between multiple networks.

As mentioned before a few times, economic theory is generally based on the assumption that information is a homogenous good with non-rival and excludable properties. This model creates a more realistic depiction of information by allowing for "near" non-rival, "near" non-excludable, and heterogeneous representations of units of information, referred to in this model as "memes." The model may consist of either two or three networks, depending on user specifications, and with each network representative a distinct language, such as biological, social, or computational. Memes can be "learned" either from the environment or through interaction with other language networks. Memes can also be forgotten over the course of the model run, allowing for a degenerative process to information retention, or they can be strengthened, allowing for the possibility for a meme to gain significance to its network over time.

There are three main goals of this model: (1) To show the difference in communication between biological, social, and computer networks. (2) To show how fundamental economic models of information can be implemented into biology. (3) To provide a model with which other researchers can explore intuitively various scenarios of communication.

4.3.2 How it Works

The memes are created based on the number of language memes and probability sliders on the interface. This first slider determines how many memes each network will possess. The second slider determines the probability that a link will form between any two memes in the same network, based on the Erdos-Renyi model for generating random graphs. Essentially, a given meme will attempt to form a link with every other meme in its network, but will only be successful if the probability is satisfied. For the purposes of visual clarity, each language network is a different color and clustered into a different quadrant of the NetLogo grid (see Fig. 4.1).

Communication between the networks: In this model, one network is designated as "O1" while the second is designated as "O2." When a third network is instantiated, it is designated as "O3." Upon setup, one of these networks is assigned as the signaler network. Whichever network is the signaler will attempt to send information while the other network will attempt to receive information. At the end of each model tick, the designation for "signaler" will switch. In effect, this allows the language networks to take turns attempting to send and receive information.

The communication method chooser on the interface determines the manner in which the networks will attempt to acquire information. A detailed description of the function of each option on this chooser can be found under Sect. 4.3.3.

The number of networks chooser allows the user to select either "two" or "three" networks to be created at setup of this model. When there are two networks, each will take turns being either the sender or the receiver. When there are three networks, the sender and receiver will be randomly selected at each tick of the model.

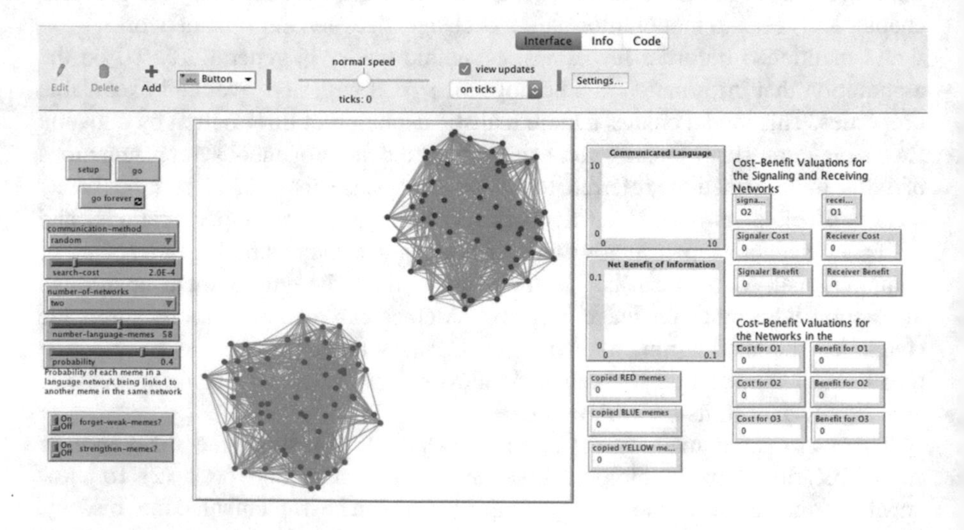

Fig. 4.1 The NetLogo interface of the agent-based model

In this model, information is exchanged in the form of subnetworks consisting of 1–10 memes each. Successful communication of a subnetwork, S1, is contingent upon a cost–benefit analysis based in economic theory. Below is a description of how the costs and benefits of communicating are calculated for the signaler and the receiver.

The cost to sender is the number of memes in the subnetwork divided by the number of memes in the sender network. The benefit to sender is the number of actual links of the subnetwork to the sender network divided by the potential links of the subnetwork to the sender network. The cost to receiver is equivalent to the value is equivalent to the search-cost slider on the interface. It is a fixed value determined by the user. The benefit to receiver is the number of potential links created between the subnetwork and the the receiver network divided by the number of potential links between the subnetwork and the sender network.

Forgetting weak memes is a simple switch to turn the degenerative process on or off for all networks. When this switch is on, the memes in each network with the least number of links will die and their links will be removed from the network. This switch represents the process by which information and language can be refined through the process of having inessential memes pruned from the network.

Strengthening memes is a simple switch to turn the process for strengthening links on or off. When this switch is on, a random set of memes will be allowed to acquire more links to their network. The number of memes that are strengthened each tick is determined by the memory retention formula.

4.3.3 How to Use it

To use this model, the user should first select the number language memes and probability values desired from the sliders on the interface. The number of networks chooser should be used to designate how many networks will be created. Then, by clicking the setup button, the appropriate formulation of language networks will be instantiated.

The user can then use the communication method chooser to determine how the two networks will attempt to exchange information memes. It is possible to adjust this chooser for each tick of the model by only clicking the go button once. However, to choose one fixed method of communication for the entire model run, the user should use the go forever button. There are four possible selections under the communication-method chooser; respective functions are as follows:

- Random: Choosing this option will allow each language network to randomly choose between one of the three communication methods below for its turn.
- Send and-receive: This option will result in the network designated as the signaler to randomly choose a subnetwork of memes (1–10 memes total) from its network and attempt to send it to the network designated as the receiver to be incorporated into the latter's network. The subnetwork will only be successfully

communicated if both language networks find the benefit of communicating to exceed the cost.

- Environmental-learning: This option allows whichever language network is currently designated as the receiving network to acquire a subnetwork of memes from the environment. When this option is selected, a random collection of memes is created in the environment (1–10 memes total) and which will attempt to join the receiving network. This is only successful, however, if the benefit of adding these memes to the receiver network is greater than the cost.
- Selective: This option represents a language network with the ability to selectively choose to incorporate either a subnetwork from the signaler or a subnetwork randomly created from the environment. To do this, the benefit of the signaled subnetwork is compared to the benefit of the randomly created subnetwork. Whichever has the greater benefit will be incorporated into the receiving network as long as the cost is still outweighed.

The two graphs, Communicated language and Net Benefit of information, allow the user to track the results of each model run and compare the results under various conditions.

4.4 Communication Scenarios and Measurements

Future versions of this model might implement more than three language networks. The significance of this expansion would be to investigate how communication changes as the number of information networks increases.

Another future version of this model might allow multiple language networks to communicate information simultaneously, with only the most beneficial subnetworks being successfully communicated into receiving networks. This expansion might reveal insight into the methods of communication between groups, rather than just binary exchanges of information.

Here are a few case studies based on the behavior of the model relating to different communication scenarios.

4.4.1 Cases in Biological Communication

1. Keeping the communication method at random, reducing the search-cost to 0.005, number of networks to 3, language memes to 100, probability 0.5, forget weak memes turned on, and turn off strengthen memes. The model shows natural selection at work, and how species can have unfavorable characteristics in certain environments and are unable to compete, and eventually lead to extinction (this is the purpose of forget weak memes being on).

2. When you turn off forget weak memes and strengthen memes, and you put the probability to 0.3, and you put the search-cost at 0.001, have the communication on random with three networks and 100 number language memes. Having this setup allows for one network to always completely surpass the other networks. However, the other networks do not totally die off. This is similar to how natural selection works on two different traits before there is a divergence in species. This time, it is not that the animals with unfavorable characteristics simply die off. Rather, this can show how the species diverge with different characteristics over time. They continue to proliferate (one may be proliferate more), but they do not compete.
3. Having the communication method as send-and-receive, turning off forget weak memes and strengthen memes, search cost at 0.001, three networks, 100 language memes, and a probability of 0.1 can show the randomness of diffusion. Often times after pressing "go," nothing may occur but energy will be spent. In the same way, facilitated diffusion is random in the sense that energy will be spent to bring the molecule across a gradient, but the individual molecule has random movement, and can move anywhere—even away from target destination.

4.4.2 Cases in Human Languages

1. The "base" case for how humans communicate with each other can be modeled by selecting two networks, send-and-receive communication, the forgetting-weak-memes switch off, and the strengthen-memes switch on. Under these configurations, both networks grow rapidly as they encounter and incorporate memes from the other network since both are actively attempting to communicate with each other. Both networks also end up with a roughly similar amount of acquired memes, a sign that the communication has been mutually beneficial.
2. Another type of human communication occurs when the two people in question to not share the same language. In this case, although both may be attempting to communicate, it is still difficult to convey meaning. This can be modeled by selecting two networks, environmental-learning, the forgetting-weak-memes switch off, the strengthen-memes switch on, and a high search-cost. The high search-cost represents greater difficulty in extracting meaning from the environment, as one might feel in a country that speaks a foreign language. However, human communication in an environment in which the language is understood can be modeled by selecting two networks, selective communication, both the forgetting-weak-memes and the strengthen-memes switch off, and a low search-cost. In this case, the two language networks have a choice between either extracting meaning from the world around them or learning from each other. With the search-cost low, we can represent an environment in which the person does not have to struggle to understand (i.e., they are fluent in the language).
3. Another interesting case occurs if we set the model to run with two networks, selective communication, the forgetting-weak-memes switch off, the strengthen-

memes switch on, and a low search-cost, but only run this model for 10 ticks! At this point, one of the networks may have acquired more new memes than the other. By then switching the communication method to environmental-learning and clicking the go-forever button, we see a huge increase in the number of memes that the networks acquire. This represents an agent that is fluent in one language (the selective communication stage), but then is suddenly exposed to a new language (environmental-learning). At this point, the agent's language network grows rapidly, just as can be seen when a person is immersed in a foreign language. Human usually learn a new language most efficiently if dropped right in and forced to adapt, as can be seen in this model. In addition, by being exposed to a new language, human knowledge grows expansively over a brief period of time.

4.4.3 Cases in Computer Science Communication

1. The agent based model can be used to demonstrate different types of machine learning techniques. For all of these algorithms, the forget memes switch must be turned off and the strengthen memes switch must be turned on since we are dealing with computers here. The supervised learning technique in which the computer is presented with example inputs and their desired outputs known as the training data can be modeled by using two networks and selecting the send and receive communication option because the computer learns from only the data that is fed to it. The probability slider can be a good measure of data size variation since more the information fed to the computer, higher is the probability of number of links within the example input. The size of the training data set can be controlled by sliding the probability pointer which demonstrates that the learning curve grows faster with more accuracy (both networks converge throughout the curve) as we go on increasing the training data set size. The unsupervised learning technique wherein no labels are given to the learning algorithm, leaving it on its own to find structure in its input can be modeled by selecting the environmental learning option and keeping the other settings same as above. The resulting curve is diverging can be used to determine data patterns and groupings.
2. Another case will be considering the memes as information packets in the internet connections by using the send and receive method. We can demonstrate the case of multiple hosts on the internet by selecting three networks in our model. Strengthen memes switch should be turned on to account for packet re-transmissions and duplication in TCP (Transmission Control Protocol) type connection oriented communication and the forget memes switch should be turned on to account for packet drops and losses in the routing paths while demonstrating UDP (User Datagram Protocol) type non-guaranteed datagram delivery. The results show that using the settings for TCP protocol, the communication between the client and server is slow (more ticks required) though most of

the information is exchanged accurately, whereas in the UDP case, the three host curves are highly diverging showing loss of information and inaccuracy but the communication has low latency as the communication takes less time in terms of ticks. It can also be seen that the net benefit of information is much higher in TCP than in UDP which verifies that TCP connections are more reliable. This provides an excellent demonstration of the differences between TCP and UDP transport layer protocols.

3. The dynamics of use of different programming languages can be modeled choosing a very high search cost justifying the fact that it is very costly to migrate software applications and infrastructure from one language to another. In any run, one of the languages shows swiftly and extensively increasing trend while the weak languages either reach a stagnant state or slower growth. This explains that at any given time, the most popularly used programming languages keep gaining even higher popularity consistently showing a large gap between the older sparsely used coding languages.

4.4.4 Some Interpretations and Overview of the Model

As I have mentioned before, Shannon is well known for laying out the foundations of information theory several decades ago. Every model of information diffusion and information exchange has essentially followed Shannon's theory of quantifying information and information exchange as sender-message-receiver. But in the natural and social world, information is essentially characterized not only by quantity, but also by subjectivity, individual interpretation and meaning, characteristics that have been largely left aside from the information diffusion models, including many social and biological models of information exchange. This research revisits Shannon's theory and shows how we can model and quantify subjectivity and meaning in information by using principles from the economics of information. It also shows how, using an economic approach, we can model internal adaptation and selection communication mechanisms both for the sender and receiver. We show that these mechanisms, under specific conditions that replicate physical and informational/cognitive decays in DNA, lead to communication emergence. The simulation results are validated with previously published laboratory experimental results of the cAMP gene signaling effects, gene that is responsible for signaling in cell morphogenesis. Therefore our model, or at least our economic approach to modeling signaling and communication, can help molecular biologists advance their understanding of signaling effects in morphogenesis or signaling effects in gene sequencing. This model can also advance the new developments in the field of DNA digital storage.

The literature that ties economic analysis with communication in the biological world is sparse. An interesting economic model of animal communication is described in Bradbury and Vehrencamp [13], which shows the dependence of the cost and benefit of information exchange between animals. On the contrary,

Table 4.1 The parameters of the agent-based model

Parameters and parameter values in simulation	
No. memes	$n \in [1, 100]$
No. networks	$N \in \{2, 3\}$
Prob. connection	$p \in [0, 0.5]$
Search cost	$c \in [0, 0.01]$
Memes decay (τ)	on/off
Memory retention (R)	on/off

work by Jakob Bro-Joergensen [15] shows that animal communication cannot be viewed only as an economic process. It is argued that animal signaling depends on environmental fluctuations and signals may be repeated in spite of the cost/benefit of repetitive signaling. An influential study by Pollard [68] exemplifies this evolution of communicative complexity in animals by taking specific case study of sciurid rodents.

In the model we developed, units of information are referred to as "memes" [22] and are represented as nodes in a network. Each meme is linked to others in the same network based on the Erdos-Renyi model for generating random graphs [27]. In this way, all memes have a homogenous probability of being linked once to every other meme in its network. The user can control this standard probability using the slider on the interface of the model. To aid in the visualization of these two networks, each colored differently (red for organism or network O1, blue for organism or network O1, and yellow for organism or network O3, in the case of three networks) and memes of a common network are clustered together on a quadrant of the NetLogo grid. The parameters of the model are presented in Table 4.1.

With these simulations, we are tracking the emergence of communication in pairwise interactions between cells or organisms using a computational, agent-based model. We are using economic concepts of information exchange, network topologies, physical decay, memory retention in a computational model of pairwise interactions. We are looking at four different scenarios of communication: cell–cell (organism–organism), cell (organism)–environment, random between cell (organism)–environment and selective between cell (organism)–cell (organism). Our agent-based model shows that with any identical two cells or organisms that communicate, the "selective" method of communication leads to more communication and more information exchange in the system (see Fig. 4.3). The selective method of communication is based on a choice of communicating with the other cell (organism) or with the environment, given the highest expected net benefit of communication. Our model also shows that cells (organisms) that initiate communication are the ones that survive the longest and that develop larger ontologies.

These results hold even when we scale up from two to three cells (organisms) and are independent of initial conditions and not sensitive to any parametric values. In essence, we show that a simple theoretical variation of the state-of-the-art models of communication (instead of focusing on the sender-message-receiver framework, we focus on the heterogenous process of information exchange) makes possible the detection of emergent signaling/communication.

We are validating our model with data on cAMP, which is the molecule that controls for signal aggregation, hence communication; as cells aggregate, high, continuous levels of extracellular cAMP functioning through the cAMP receptors activate a transcriptional cascade that leads to cell-type differentiation and morphogenesis. Cell movement and cell-type-specific gene expression during development are regulated by cAMP, which functions both as an extracellular hormone-like signal and an intracellular second messenger. We therefore suspect there is a close link between pairwise information exchange, as shown in our model, and morphogenesis and that the activation order of genes in morphogenesis is highly dependent on their communication.

The main goal of the computational model was to detect emergence from attributes and behavior rules and not from initial conditions, given these simple, economically altered rules of information exchange.

The analysis of the data from the computational model shows that implementing economic concepts of benefit and cost of communication that are heterogeneous for identical/informationally similar cells or organisms leads to one method of communication (selective described above) to be responsible for the most interesting behavior in the system. These results hold even when we scale up from two to three cells (organisms) and are independent of initial conditions or any other parametric (search cost, number of nodes, etc.) values.

The data from the model includes 16 combinations of scenarios (4 communication methods, forget and memory on/off), using a full parameter sweep for the probability and search-cost parameters in each of these scenarios.

Communication "happens" in 97% of the attempted simulation cases when the agents communicate only with the environment, in 65% of the cases when they communicate only with each other, in 60% when there is random choice between environment and peer communication and only in 32% of the cases when there is selective communication activated.

4.4.4.1 Power Law

The first check was whether the Zipf distribution holds for the communication produced and exchanged in these simulations [83, 100]. Only the selective scenario of communication with both memory strenghening and physical decay (on/on switches) yields a power law distribution, therefore closer to the natural language, while the other scenarios do not (see Fig. 4.2).

The data with respect to the *value* of the nodes (the net benefit of communication at each time step) was binned in 100 bins and plotted on a log-log plot of rank and size.

$$p(x) = Cx^{-\alpha}, \tag{4.1}$$

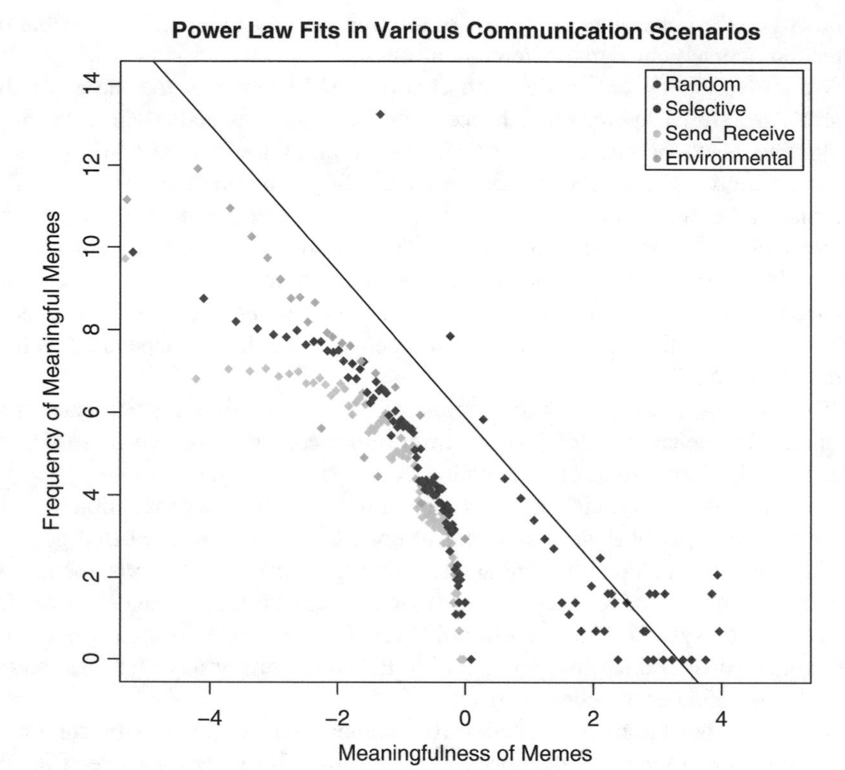

Fig. 4.2 Power Law distribution of communication in each scenario. In a log/log plot where the data was binned in 100 bins, the selective method of communication with on/on switches shows a power law distribution of communicated memes, while the other scenarios do not

where $p(x)$ is the probability distribution of x, $x(t) = \delta B(t) - \delta C(t)$ individual net benefit of communication at each time step; $\alpha = 2.36$ and $x_{\min} = 0.12$.

The power law tests for the actual number of memes/nodes exchanged at each time step did not show any power law distribution closeness with respect to the actual number of memes. While power law distribution was found for the values of the memes, one inference that can be made is that the "meaning"/value of information is potentially one of the drivers of the Zipf regularities in natural, organically developed communication.

4.4.4.2 Entropy

In these models, communication is being produced and exchanged with the same probabilities. Therefore a true Shannon's entropy measure would be irrelevant for each network or scenario of communication—the full parametric sweep records data on all range of probabilities. The best proxy for information event probability we can

Table 4.2 Average entropy values for each scenario

Scenario:	Selective	Random	Network–Envir.	Network–Network
Average entropy	0.63	0.01	0.14	0.11

calculate with any meaningfulness is the measure of the net benefit per observation (x from above) relative to the network probability of connections (density):

$$H(t) = -[\delta B(t) - \delta C(t)]/p(t) \tag{4.2}$$

where p is the parameter in Table 4.1.

The values of the entropy for each of the four scenarios are given below (see Table 4.2).

4.4.4.3 Quantitative Communication

Figure 4.3 shows the communication results over the entire parameter sweep. The selective communication scenario is the one that has the highest means and variances of communication both in the number of memes communicated and in the number of memes/nodes in their networks.

4.4.4.4 Value of Communication

As the focus of these models has been on the value of communication, the net benefit of communication is also singled out as the highest means and value range and for the selective method of communication (see Fig. 4.4).

Therefore the selective method of communication is the most distinctive one for both the amount and the value of communication exchanged in this system.

4.4.4.5 Life Time of Communication

The selective communication also shows a slightly higher mean and range of life span for the communicating networks than other scenarios (see Fig. 4.5). This is quite an interesting result, as it shows how information helps offset slightly the physical decay of the networks, which is constant for all networks in the system.

Another result that we also observed is that forgetting is stronger in effect than memory strengthening.

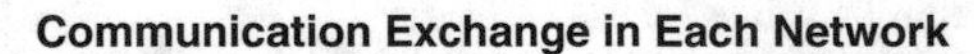

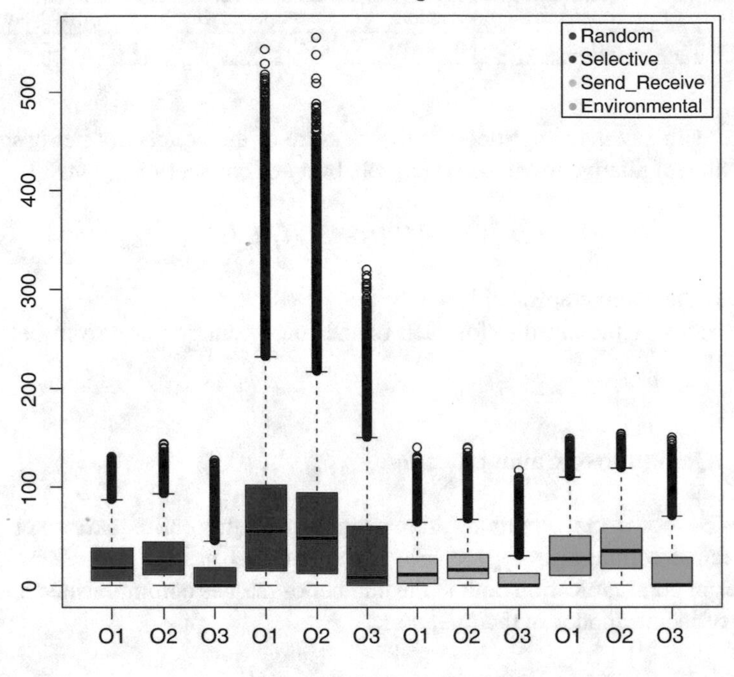

Ontology in Each Network

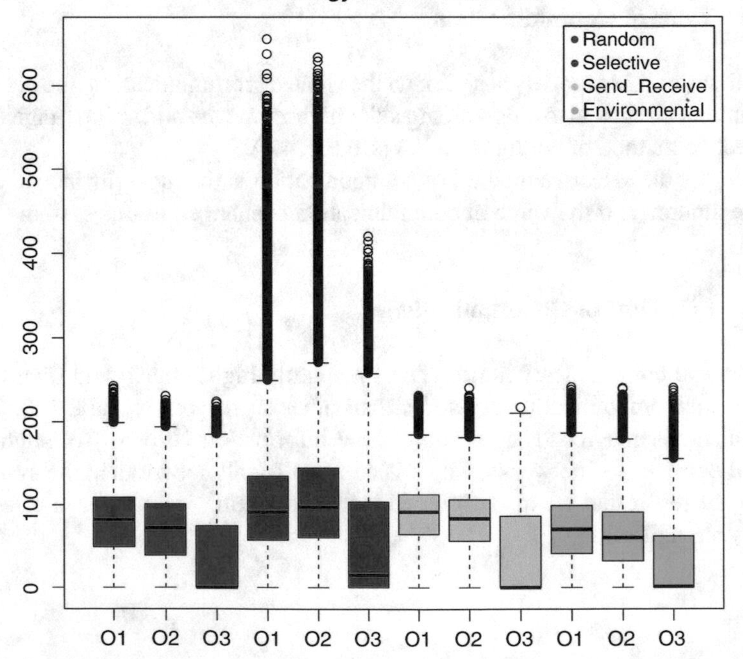

Fig. 4.3 Individual communication in each scenario. The selective method of communication shows the highest mean, median and variance than any other method of communication for each network

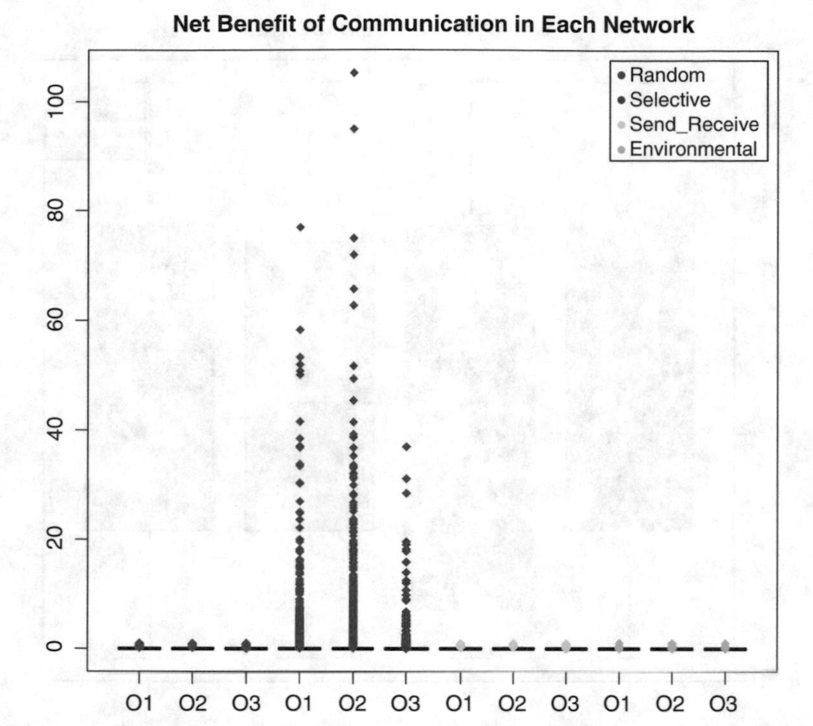

Fig. 4.4 Individual communication in each scenario. The selective method of communication shows the highest mean, median and variance than any other method of communication for the value of communication

4.4.4.6 Sensitivity to Initial Conditions

The model was also slightly adapted so that we can control for which network starts communicating first and re-did the parameter sweeps with either network O1 and O2 as the initial signalers and we found that the model results presented above are robust to initial conditions. These results hold in three networks scenarios too.

This model is using the widely accepted computational technique of agent-based modeling in social science and economics. It shows that a simple theoretical variation of the state-of-the-art models of communication (instead of focusing on the sender—message—receiver framework, we focus on the heterogenous process of information exchange) makes possible the detection of emergent signaling/communication sequentially.

While the analytical model sets a simple graph based theory of communication by incorporating quantitative measures of value borrowed from economics in heterogeneous networks, the computational model explores various scenarios of communication where information exchange is repeated and subject to a few extra parameters and settings that would bring the information exchange model closer to real communication in biological and social networks.

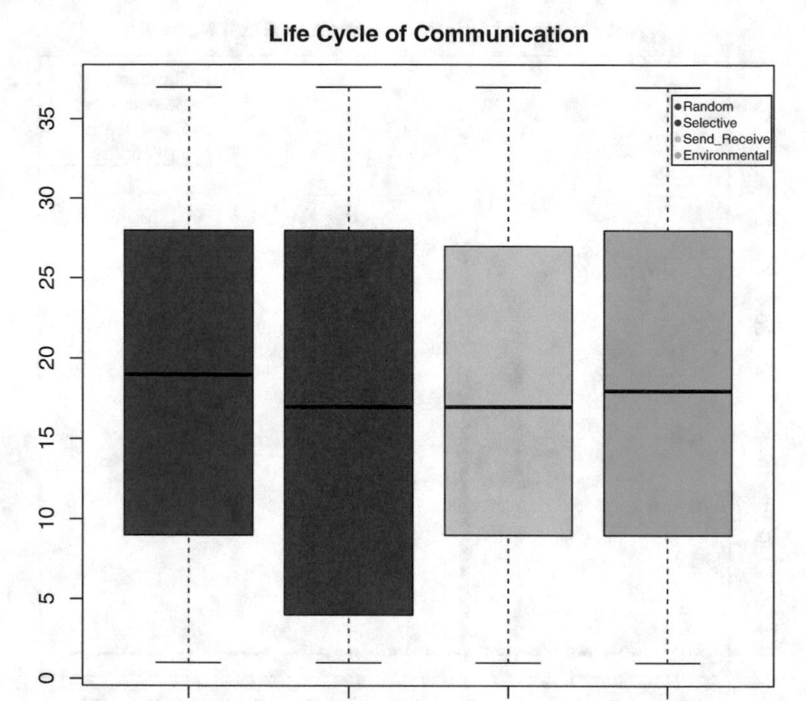

Fig. 4.5 Individual communication in each scenario. The selective method of communication shows the highest mean, median and variance than any other method of communication for the value of communication

The computational model shows that given any identical two networks that communicate, the selective method of communication leads to more communication and more information exchange in this system. This selective method of communication is based on a *choice* of communication between the other network and the environment, given the highest expected net benefit of communication. This means that networks (organisms) that have the ability to change the rules by which they interact develop informationally differently from networks (organisms) that only react to stimuli. And it also shows that overall this type of communication is more beneficial for the system of networks (organisms) overall, making communication a collective good.

These two models show that implementing economic concepts of benefit and cost of communication that are heterogeneous for identical cells (organisms) and simultaneously implementing physical decay and memory (information) retention leads to one method of communication (selective described above) to be responsible for the most quantity of information exchanged in a system and retained by the cells (organisms). These results hold even when we scale up from two to three cells (organisms) and are independent of initial conditions or any parametric values. Therefore a simple theoretical variation of the state-of-the-art models of communication (instead of focusing on the sender—message—receiver framework, we focus on the heterogenous process of information exchange) makes possible the detection of emergent signaling/communication sequentially.

The problem of biological and social communication and the findings of our models are relevant to the immunology and genetics communities regarding the signaling sequence and gene activation in cells, as well as to trust and reputation networks research in social sciences.

Future work on this new approach to information exchange includes computational models of other network topologies, as well as scenarios of communication when the pairwise communication is attempted between networks that are originally informationally different.

Chapter 5
Intra Versus Interspecies Communication: Boundaries and Advances

5.1 Boundaries in Communication

I believe we are moving towards a more integrated science and view of behavior in social and biological systems, where we are less focused on the boundaries between the systems and sciences, but are rather tracking and studying behaviors across systems and species. For example, we are more likely to get insights into the origins and evolution of emotions when we are look not only inside our brain or within our understanding of cognitive and psychological sciences, but when if we try to determine rudimentary or basic emotions in other species, as lower as we could possibly get down on the evolutionary scale. For example, some of the latest research on the sentience of various animals shows that emotions can be detected even in insects such as bees [6] and some controversial research mentions even plants as having some basic emotions such as fear or pleasure.

The tracking of a process, phenomenon, or behavior across all spectrum of life possible (or, better said, across the entire taxonomy) is quite exciting and potentially more revealing of our human behavior than studying ourselves alone.

A very good question is whether the gender can alter the information exchange both in animals and in humans. At large scale, matrilineal transmission of information and culture is different from the patrilineal transmission of information and culture. In the matrilineal transmission of culture, most females spend their lives in the same social group as their mother while both are alive and the daughters stay with their mothers. This is not the same with "matriarchal," which means female elders have power or influence. Elephant societies are both matrilineal and matriarchal, while the large toothed whales live in matrilineal societies [94].

Another boundary of communication is given by the environment or the ecology where the individual and the group live. Cultural transmission can cause variation in behavior within a species [94]. For example, environments lead not only to different body types (morpho-types) in species, but also to groups with different ways of life (eco-types).

A. Berea, *Emergence of Communication in Socio-Biological Networks*,
Computational Social Sciences, https://doi.org/10.1007/978-3-319-64565-0_5

> "It is not enough though simply to demonstrate that animals in one place, or one group, behave differently to those in other places or groups. Environmental variation can also produce differences in behavior between communities in several ways. For instance, if animals can learn a behavior with no social input, then in places with a particular food or tool present these may get used, whereas their use will be absent from communities where the necessary food or tool does not occur. This behavioral variation is not culture; it results from individual learning in different ways on their own. Genetic variation can also lead to differences in behavior" [94].

5.2 The Story of "Whale 52": The Loneliest Whale on the Planet

As humans, we do like to reference the animal world (the "other" animal world) to our own and draw inferences from our own types and ways of communication and extrapolate it to other species. But is this an accurate way—not even accurate—a "good" enough way to look at the communication between other species?

Bill Watkins, a marine mammal researcher at the Woods Hole Oceanographic institution (WHOI) in Massachusetts tracked the songs of a particular whale and the geographical locations for more than a decade (12 years to be more exact) and, due to the fact that the whale seems to "sings" out of the patterns of any other species of whales, popular media has named it "the loneliest whale in the world."

Some later research points to the fact that this might not be the case, that whales change their calls and that it is not that uncommon for marine mammals to sing out of their patterns.

The story of the "whale 52" has fascinated mass-media and popular science. There is currently even a documentary in the making about the story of the whale, a crowdfunding campaign (both for the documentary and to find the whale) and many articles featured in BBC, Discovery Channel and Smithsonian. Why would a story of a whale that is "out of its pattern" fascinate us so much?

The absence of communication can be not only painful, but also life threatening, as it is a signal of the absence of any peers, any social environment, and any opportunity of growth and benefit. Similarly, the presence of first communication signals contact and the presence of a peer that we can grow with, explore with, and benefit from. Perhaps that is why stories like "Hello world," which signals our first contact with computers and our first communication in a digital world, or of the Voyager and Pioneer plaques are so touching and have become social norms and symbols. And our mournings (and animals mourn too, such as elephants, monkeys, and dolphins) of the lost ones are yet another example of the loss of communication that is permanent.

On the other hand, both humans and some animal species can display "autism," which is a different form of communication. Due to the differences in social behaviors and social interpretations of communication, autism is a condition that shows how close social behavior and communication are both in humans and animals. Some experiments on rats showed how difficult it is to understand social behavior in animals without the ability to properly measure it and how important it is for scientists to pursue models and social behavior and communication together [99].

5.3 Computational Semiotics and Biosemiotics Models

"Biosemiotics can be defined as the science of signs in living systems" [51]. Biosemiotics is an interdisciplinary field built on the idea that there are meanings and interpretations at the cellular and molecular levels and that there are interpretable signs in the organic code that should point towards some deeper, more fundamental underlying informational processes that drive life itself and all organisms [4]. One of the core postulates of biosemiotics is that the cells and proteins are artifacts and not spontaneous molecules, due to the ribotype that is responsible for the genetic code according to some natural laws that we don't currently fully understand.

> There are, in conclusion, three major differences between copying and coding: (1) copying modifies existing objects whereas coding brings new objects into existence, (2) copying acts on individual objects whereas coding acts on collective rules, and (3) copying is about biological information whereas coding is about biological meaning. Copying and coding, in short, are profoundly different mechanisms of molecular change, and this tells us that natural selection and natural conventions are two distinct mechanisms of evolutionary change [4].

Semioticians and biosemioticians have been pointing towards the idea that the origins of life itself is based on signs and meanings for a long time and that there is a universal code that we can potentially, sometime in the future, understand and compute in order to get a deeper understanding of these fundamental processes of information and meaning of information [26].

An interesting approach to the development of algorithms and computational models from semiotics has been done by Gomes et al., who took the principles of Piercian semiotics and his fundamental model and developed a simple algorithm to simulate the emergence of "artificial semiosis" [31]. This algorithm was not developed or implemented into practice or a computer model, but computational semiotics has become a field of interest during these past few years for the development of artificial intelligence. At the core, computational semiotics deals with autonomous intelligent systems that perform any kind of intelligent behavior, including perception, world modeling, value judgement, and other types of behavior.

Computational semiotics is an interdisciplinary field between semiotics and computational sciences that is particularly focused on "autonomous intelligent systems able to perform intelligent behavior," which includes perception, world modeling, value judgement and behavior generation. There is a claim that most part of intelligent behavior should be due to semiotic processing within autonomous systems, in the sense that an intelligent system should be comparable to a semiotic system" [35]. Computational semiotics has great promise with respect not only to advancements in artificial intelligence, but with respect to advancements for a general theory of communication as well.

Vigo presents an interesting formalization of the "Representational Information Theory and Generalized Representational information Theory" as a way to account for informational content from the "meaningfulness" paradigm instead of just the quantitative paradigm that is prevalent in computer science and computational models of information [91].

An unconventional language for a long time was the one given by the symbolism of flowers and flower arrangements. As it is the case with any other secret languages or informal symbols that have been deeply embedded in our cultural history, flowers and flower arrangements were able to capture a secret language to depict emotions that otherwise would be too difficult to describe at the time [76]. The classic example is that of the red roses to express love or of yellow flowers to express jealousy. In many ways, flowers were the "emoticons" of an era before our digital times. Japanese culture elevated flower symbolism to an art form, named "ikebana".

Animals don't have normative language and cannot ask "Why" questions or be aware of knowledge they don't know. They show a wide range of emotions and empathies similar to ours, but when it comes to meta-cognition and normative understanding, animals are very limited. Therefore a semiotic analysis of animal signals and interpretations is strictly linked to a biological interpretation of signals and symbols, but in humans, just like in the example with the flower symbolism, semiotics evolves with the new language constructs or means of communication that we assign to messages. In other words, the balance of information exchange shifts towards the means of communication and their symbolism much more than to the message itself.

In computer science and modern information theory, some of the newer methods to incorporate semiotics and meaning in information exchange include developments in fuzzy logic and ontological developments and applications [52].

In our agent-based model described above, the meaningfulness of information is quantified as the net benefit of information for each network as well as for each information exchange (see Fig. 5.1).

5.4 Miscommunication and Deception from Cellular to Organisms to Human Communication

As we have seen so far, communication is about the exchange of information between any two organisms or living forms, but it does not address what type of communication is being transmitted. For us, humans, we are quite aware of the fact that the information exchanged might not always be the best or the most honest. But miscommunication or deception are not human traits. Very often, communication and miscommunication (purposefully or not) are two sides of the same coin and one relevant question that we can ask is: Do animals lie?

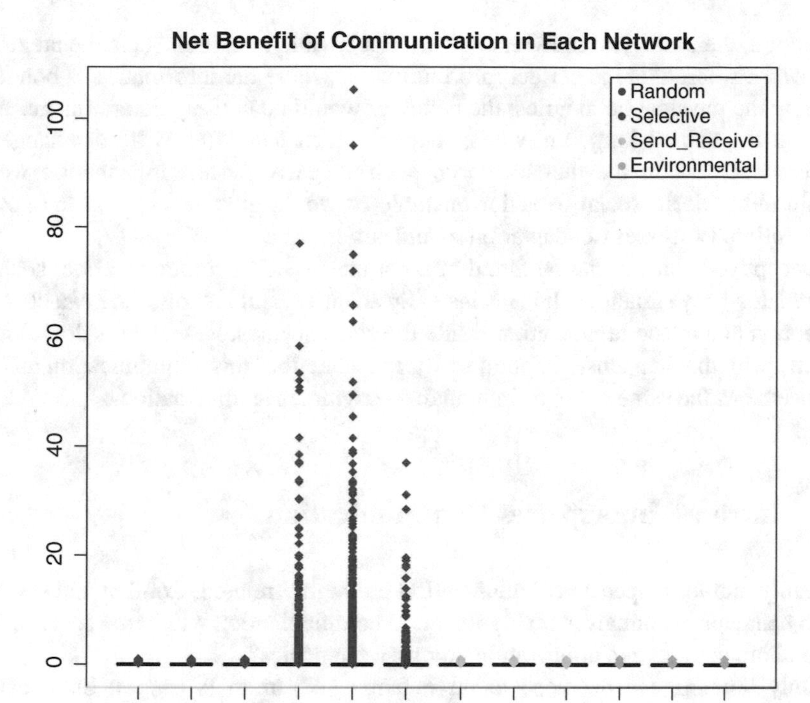

Fig. 5.1 The net value of communication in each scenario from the agent-based model

Just as communication starts cellular-at the level of the cells-so does miscommunication. Miscommunication between genes is called "intragenomic conflict" and it is believed to be one of those fundamental evolutionary processes that are at the basis of natural selection and perhaps even at the basis of the evolution of sexes. Intragenomic conflict happens when one gene transmits genetic information under different rules than the other genes or when the gene is transmitting genetic information to the detriment of the genome. This process/effect is also believed to be at the basis of genetic/environmental (phenotypic) interaction [21].

Deceptive behavior is more common in species that cooperate more [74]. While the scientists have found lying and deception to be an evolutionary behavior that allows individuals to form coalitions and be successful in mating and food, I believe that the emergence of deceptive behavior in the biological world has a much deeper meaning than this. Deception or lying ultimately means that the sender of information or that the signaler is purposefully sending the deceptive signal in order to gain individually from the cooperative behavior of the others in their group. In a way, the deceptive signals function as a valuable way for individuals to intentionally get into a bounded rationality situation where such situation might not exist. Without deception, the biological individual would not be able to exhibit social behavior. In a network where there is perfect information, or where the information is bounded only by the physical boundaries, the behavior would be entirely deterministic. And even in the animal kingdom, where there is much less "free will" or change of preferences than in the human world, such a nearly perfect information world would either deem social behavior unstable or would give genetics all the power of selection. Processes of adaptation would suffer.

Deceptive behavior can be found in most inter-species communication, such as in predator–prey situations. It is intrinsically adaptive. But less often we can observe deception and miscommunication inside the same species. Researchers don't know exactly why this happens, although many speculate that miscommunication among the peers has the same cause as in humans—asymmetric information.

5.5 Intra vs. Interspecies Communication

Sociality and intra-species communication are highly related. Conflict and cooperation behavior in animals is highly linked to communication within the species. But what about conflict and cooperation among interspecies?

Only humans can develop "creative language"; animals cannot; but creative language can be taught to some species of animals, like birds and apes, therefore creating the potential for interspecies communication.

Interspecies communication has been a "holy grail" in interdisciplinary research. One of the longest research projects in the history of our civilization, the SETI project, has dedicated years and entire research teams to the study of interspecies communication with the purpose of gaining insights into the search for extraterrestrial intelligence. Interspecies communication has been a somewhat esoteric field of research as well in communication, as many research scientists studying it have been viewed as being on the fringe of their discipline.

In nature, interspecies communication has been observed in several species: at the cellular level; in primates, in dolphins and dogs and of course in humans and dogs or humans and horses or other animals.

But perhaps the species that is the most studied and researched both with respect to communication and with respect to social behavior are the ants, who also gave rise to a new concept in social behavior, named "eusociality."

Elephants, like the marine mammals and the apes, are able and capable of complex communication patterns and even interspecies communication. Accounts of elephants mimicking human voices, technological sounds (such as truck sounds) have been recorded and made an impressive impact into science popularization. The website of the organization "Elephant Voices" reports that like humans, are able to make and use tools, and show evidence of social learning, advanced acoustic, visual, chemical and tactile communication, are able to communicate and maintain contact with other elephants over long distances using seismic communication signals, which they absorb through their feet, can discriminate between the voices of at least 100 other elephants. Their basic social unit is the family, which includes a mother and her sexually immature offspring, they live in complex societies, that separate and reunite based on weather conditions and food availability, their families are led by matriarchs, who store decades of ecological knowledge that is critical for the survival of the family unit and members through droughts, predation, and other threats. Elephants also tend to have lifelong or long-lasting social bonds, demonstrate socio-emotional complexities, such as empathy and self-recognition, and display concern for distressed and dying elephants [25].

Like all highly social mammals elephants have a well-developed system of communication that makes use of all of their senses—hearing, smell, vision and touch—including an exceptional ability to detect vibrations. At one end of the spectrum elephants communicate by rubbing their bodies against one another, at the other end they may respond by moving toward the sounds of other elephants calling, perhaps 10 km away [25]. They convey information about their physiological (e.g., sexual/hormonal, body condition, identity) and emotional state (e.g., whether they are fearful, playful, joyful, angry, excited) as well as communicating specific "statements" about their intentions or desires.

Perhaps the most interesting of all types of complex communications in elephants is the seismic communication. According to the same website, seismic energy transmits most efficiently between the 10 and 40 Hz-in the same range as the fundamental frequency and second harmonic of an elephant rumble [25]. When an elephant rumbles, a replica of the sound is also being transmitted through the ground.

Another group of animals with very interesting and complex means of communication both intra and interspecies are the cetaceans (whales and dolphins). They communicate with coda which are morse-like pattern of clicks, through movements of the fins striking the water and through eco-location.

The fishing cooperatives are extraordinary interactions between wild dolphins and humans. Perhaps the only other example of such cooperation that does not involve capture or training is the way that greater honeyguides-small birds that live in sub-Saharan Africa-lead humans to bee colonies using special double noted calls and posture signals that indicate where the hive is [94].

The ideas of swarms/swarm intelligence and emergence are closely linked with each other. Swarms have been observed in biology as a form of collective behavior and movement (in birds, insects and fish in particular) and have been some of the first forms of behavior simulated using genetic algorithms or agent-based modeling. What makes swarm behavior particularly interesting is the fact that there is no central design of the behavior, no leader to direct the crowd, the behavior is highly coordinated and synchronous and yet there is no clear understanding regarding the purpose of it. Unlike swarms, stygmergy is a form of self-organization that involves indirect communication and coordination. In a way, stygmergy is also a form of communication and coordination, as well as collective behavior, with the caveat that the signals or the transmission of information is performed indirectly and thus can be lagged or spatially bounded. One good example of stygmergy is the communication through pheromones. And the current robotic and multi-agent systems algorithms are trying to model stygmergy even more than modeling swarms in order to optimize their collective behavior.

The word "intelligence" in swarm intelligence is particularly misleading. It points towards the fact that, although each individual agent follows very simple rules of behavior and acts based only on the local and bounded environment that is accessible to it, the global, emergent behavior is similar to the one of one entity acting with a certain purpose. Perhaps one of the most archaic and well-known examples of swarm behavior is the proverbial locusts invasions. Swarms act as an individual either in search of food (locusts, ant colonies, bees) or to avoid predators (fish schools, birds) and the swarm is generally speaking better off for each individual than any individual could be by itself. That is why the collective action from simple rules is named "intelligence." In other words, in swarms, the collective can provide better food or defense mechanisms than an individual could on its own.

The problem of collective action in swarms is one of the greatest puzzles in nature. We currently don't know how this behavior forms, although we have pretty good algorithms that are replicating it. And the second greatest puzzle about this biological-social phenomenon is surrounding the communication of individuals that are forming the swarms.

At the other end of the spectrum are cases of social isolations, as we saw with respect to whale no. 52. The case of feral or wild children—that is of children that grew up isolated from society or from human contact, and raised by animals in the wild—constitutes a very interesting view on human communication. The research on feral children is sparse and many of the records of feral children are doubtful. Nevertheless, while some of these cases might have been hoaxes or given the very few number of cases of feral children, this phenomenon sheds some useful insights into the acquisition of language and human communication as well as on interspecies communication in the wild. The cases of wild children throughout the world are mostly tragic—the found children have very difficult time integrating back in the society, learning how to speak and write and, in general, remain behaviorally impaired to function normally among humans [53]. The sparse research on wild children tends to be focused on the problem of "critical phase" of development during childhood for humans to acquire language. There is virtually no research about how humans have reestablished contact and communication with them, no matter how little this is. But the current findings, that everyone seem to agree on, point to the idea that children need to be integrated in the social and cultural networks of peers for communication to develop, while the development of the physical brain needs to be fully integrated into these networks as well. This means that there is a physical basis for communication as much as there is a social, cultural one. Moreover, we can infer from these cases that human, natural languages are particularly social constructs and that assimilation into social networks requires more than just interactions with the peers. It also points to the similarity of human languages from these meta-cognitive aspects—there is no difference in the difficulty to re-adapt in any of the countries that reported cases of feral children.

While this is the case of social isolation in humans, in the case of animals, isolation has detrimental biological effects [16]. The authors showed that social isolations decrease lifespan in fruit fly, increase infarct size, increase stress and stress response in rats, decrease the likelihood of recovery and survival after trauma and, in general, in many types of animals studied (rats, mice, rabbits, pigs), social isolation has overall very bad consequences for the isolated animal.

But what if the animal communication is also evolving, even if not at the same pace with our communication, would it be able to catch up with us enough so that we can start communicating together? The dogs and cats have evolved to communicate with us, and this was done through repeated interactions, and the case of feral children often shows how we can communicate between species in early age. Now that we have more conservation efforts and more researchers working on these problems and experiments, perhaps we can get closer to breaking the interspecies communication barrier.

There can obviously be no analysis into the idea of interspecies communication without mentioning the work and the role of SETI. Some researchers are already proposing that the SETI studies should be focused on historic analogies based on the transmission of ideas within and between cultures rather than on analogies based on physical encounters between cultures [89].

Interspecies communication is not a fringe field, borderline with science-fiction. Interspecies communication exists in the biological world all the time: think of predator–prey situations; think of symbiosis; and think of bacterial communication. For example, even in humans, any autoimmune disease is essentially a case where the body is misinterpreting the information from the body and when different subsystems of the body cannot properly communicate with each other.

A clear case of interspecies communication in bacteria was reported by Bonnie Bassler, the molecular biologist who showed "quorum sensing" as means of communication in bacteria, that aligns gene expressions in a similar way in which social organisms organize collectively in groups or in swarms [5]. She showed the mechanism through which the cells communicate in order to collectively aggregate and invade another organism—through a signaling protein, called AX21. When other bacteria sense the presence of AX21, they create biofilms that are antibiotic resistant. In this way, the signal from one bacterium signals the others that there is danger present.

But more than a signal of intra-species communication, the same molecule can function as a signal for inter-species communication between different types of bacteria. In fact, Bonnie Bassler showed that there is a core molecule that serves as the common means of communication between any type of bacteria, while small variations in the chemistry of this core molecule determine the language for each different species of bacteria. In other words, there is a molecular core that functions as a signal between any type of bacteria, while small variations added to this core molecule create the different language for different species—from inter-species communication to intra-species communication [5].

5.6 Means of Communication Intra vs. Interspecies

Chemical communication is a very powerful way for cells and organisms to communicate with each other in different scenarios. Chemistry has this incredible power of combination and recombination and of including or excluding information at a very fundamental level. DNA combination, recombination, and mutation is essentially a process of chemical communication. After the "age of chemistry" in communication though, life evolved to organisms that have evolved senses that are beyond a chemical exchange: sight, hearing, seismic perception, infrared vision, and so on. The evolution of these analyzers in living organisms has enabled them to communicate in a way that did not involve any physical exchange of chemicals—communication at a distance. And the third age of communication came with culture, while the fourth age came with language. Perhaps the fifth way is the digital information exchange, which is also not as much a product of biological evolution, but a product of coevolution between the human brain and language.

The latest research on communicating with plants shows that we can send "positive" and "negative" signal s to plants and receive feedback. Examples of plant communication are through systems of fungi, through airborne chemicals, and even eavesdropping on sound waves.

There are lots of interesting cases of communication in the animal kingdom, where signaling involves both sound and visual cues, but here are two examples when the visual communication is taken to extreme—communicating with pure light (the case of bioluminescence, i.e. fireflies) as well as when communication is performed in extreme environments (communication of the extremophiles) (Fig. 5.2).

A very intriguing question is about dinosaurs or extinct species communication. In the case of extinct species—at least those ones that went extinct before we started studying nature—biologists can only infer how these communicated and which were the communication patterns. They do so based on the structure of the remains of their skeletons and by association with the species that they are closest to. For example, the dinosaurs are closely associated with birds and reptilians, therefore scientists are inferring communication patterns that are mostly similar to the ones of these species.

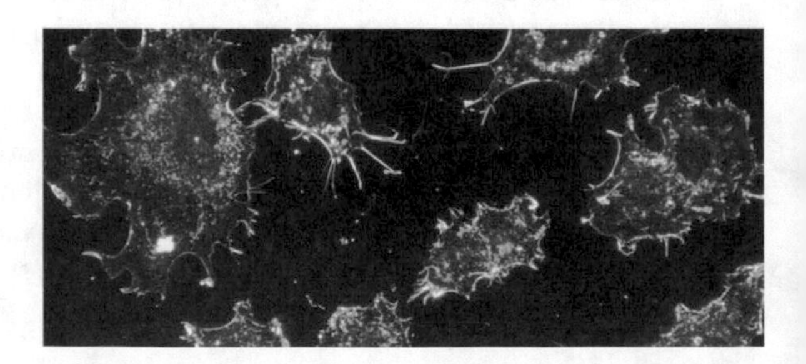

Fig. 5.2 One of the most basic forms of communication on Earth—cell to cell interaction

A very interesting way to understand how we can infer communication in the absence of evidence is by looking at the possible patterns of communication of extinct species. For example, the dinosaurs most likely communicated using low-frequency sounds that were produced in their crests that had the anatomical development that would permit them to function as sound chambers [44, 70].

Multimodal communication has evolved together with the evolution of the sensors in biology [64]. Multimodal communication is a mix of signals from different (or addressing) different sensors in the animal, with the purpose of transmitting the same message or a combination of messages simultaneously. The important findings in animal multimodal communication show that multimodal messages have the purpose not only to enhance the original message as if it were transmitted through just one channel or means of communication, but that they can produce an emergent message, which is a hybrid between the original message and the original channel of communication.

For example, researchers showed that different species of monkeys cooperate through common signals when facing a common predator [101].

Time and again studies of animal communication have identified signals that perform multiple functions, providing information along several dimensions simultaneously, information that is used in different ways by different receivers [94].

Just as in human societies, the interactions in the animal world are either cooperative (i.e. symbiosis, group hunting) or conflictual (predator–prey, competition for mating and food). Whether we are looking at individuals within the same species or individuals from different species that are interacting for one reason or another, animals, just like humans, interact either for some mutual benefit or, if such a benefit does not exist, one becomes food for another or one wins the territory or the mate over the other.

From our perspective, though, the interesting question to ask is what are the differences between communication within the same species and communication between other species, whether this is of a cooperative or of a competitive/conflictual nature. What is different in communication between other species and the prey–predator communication?

Symbiosis is perhaps the classic example of the evolution of communication between two different species (i.e., lichens). In symbiotic species, versus predator–prey communication, the means of inter-species communication are internalized and at a deeper level than the means of communication between predators and prey.

When it comes to means of communication in the animal world, one of the most interesting species to study are the bats and the way they communicate via eco-location. While many studies on bats eco-location focused on how they avoid obstacles or how they find food, very recent research into bats communication focused on the communication between themselves, in a colony [69]. This research showed that there are very distinct and different pitch calls that can be mapped as our human communication, with a sender, a receiver, a context for the message and the behavior following or preceding the call. The researchers were even able to roughly translate the calls corresponding to feeding, mating, perch, and sleep [69]. The importance of this research though comes from the fact that for the first time there have been isolated pairwise calls (instead of "broadcast" call) and that the messages have several layers of information, in another species than dolphins or primates, that have more complex "languages". Moreover, this research also connects the "language" of the bats with their social and collective behavior.

One of the most important and interesting questions that would unlock numerous possibilities both for artificial intelligence and for interspecies communication is: what is common in the way we communicate with each other as living beings? What is common in the way cells communicate with each other, in the way animals signal, in the way social groups communicate and in the way interspecies communicate with each other?

Pattee's work on cell morphogenesis showed the duality of information and medium of transmission of information: cellular morphogenesis is a great example to show how information and the means of information are interchangeable and coevolving together [65].

The first condition that needs to be fulfilled is represented by the physical capability of sending or receiving and interpreting messages. Cells would not be able to signal unless they had receptors for the signaling chemicals; animals would not be able to communicate unless they had the analyzers such as ears or eyes for sounds and light. Humans would not be able to communicate with each other unless they could talk or understand the same language.

A brand new international research project shows that the new advancements in artificial cells interacting to real, biological cells has the potential to break the barrier in communication between the silicone and the carbon worlds [54]. This research was presented as breaking the Turing test in the cellular world. The potential applications of this breakthrough both in biology, medicine but also in the information theory and computational worlds are enormous. But what is most relevant for the theory of communication is the fact that an interspecies, inter-physical worlds barrier was crossed successfully. Multimodal communication can be not only transmitting the same message to different sensors through different channels, but transmitting two different messages to different sensors at the same time—i.e. songs are an enhancement, movies are non-redundant. Particularly non-redundant multimodal messages have the power to show either new modularities (new means or channels of communication) or new messages emerging altogether.

Chapter 6
Social Versus Biological Language: The Emergence of Grammar and Meaning

6.1 The Origins of Human Language

As mentioned in the beginning of the book, economics and language are very closely intertwined. While we don't know too much about the origins of language in humans, we know a lot about the origins of writing. Writing arguably was the first artifactual innovation in human communication and it arose as an efficient means of accounting and recording crops and animals [8]. Writing is a first form of symbolic storage of information outside of the brain and outside of the oral history retained inside individuals. It also shows how the means of storing information and the possibility of *reinterpreting* it is so crucially important.

Language is also called a "living organism" and in this research we showed how by agentizing both information and the process of information exchange we can track the coevolution and adaptation between the language and its own network, and the emergence of language as a "collective" good or as a universal feature of the complex system where it was born.

On another hand, writing can also allow for information to be reinterpreted by the reader over and over again. Computers are basically taking the combinatorial essence of information to infinite possibilities.

Anthropological researchers have found that the action of pointing in humans is crucial for communication. At the Max Plank Institute, they showed how the simple gesture of pointing towards objects and then articulating what these particular objects are, leads to the development of vocabulary/(lexicon). Gestures and sequences of gestures also enabled the representation of actions and thus a proto-form of story telling in very ancient humans.

Visual communication is still the most poignant means of communication that we know and understand in the living ecosystems on this planet. Vision and conversion of 3D into 2D in our brains is one of the most wide spread complex processes but we know little about the mechanism behind it. While hearing and vocalizing played and still plays a great part into our communication and language, as well

A. Berea, *Emergence of Communication in Socio-Biological Networks*,
Computational Social Sciences, https://doi.org/10.1007/978-3-319-64565-0_6

as in other species, it is vision that allowed information that is transmitted between generations through means of recording to play a great role into the development of our civilization and the expansion of language. We first invented technologies for recording and preserving visual communication; only in the nineteenth century we were able to record and preserve audio communication and even to this day technologies like writing and reading and combining images with the sounds are the widest spread means of communication. Sight is better for understanding dynamics and evolution than sound is. Sight is also more permanently and better recorded in memory than sound; visual communication is easier to reproduce and disseminate than sound. For the sound based or oral traditions, the mechanisms of cultural transmission are essentially the same as the ones in cetaceans and other primates. But visual cultural transmission changed the game for the human race and civilization.

In animals, visual communication is not only signaling between predators and prey, but it is also a cooperative and coordination mechanism. In killer whales, the striking blue and white colors of the mammals can not only make them visible to prey, but it also functions as a coordination mechanism for their cooperative, social hunting behavior [94].

6.2 The Chomsky–Piaget Debate

The Chomsky–Piaget debate is arguably one of the most well-known debates in our intellectual history. In a nutshell, the debate originated in 1975 and it was largely concerned with the idea of the origin of language. The debate is summarized in the very influential book "Language and Learning: The Debate Between Jean Piaget and Noam Chomsky" by Massimo Piattelli–Palmarini [19] and it is one of the most famous debates in the history of science. At the core, the debate is between the idea of a "universal grammar," posited by Chomsky, versus a constructivist approach to language, as posited by Jean Piaget. In other words, the debate is whether the language is being acquired in humans due to genetics versus cognitive factors, whether language is acquired due to our innate ability as human beings or due to our interactions with the environment. Some reviewers and commentators of the debate are arguing for one or the other.

The debate has not been settled yet. Some very recent papers on the origins of language point that some very common words have common letters and sounds in many languages, but at the same time some other newest research on the origins of language debunk more and more the idea of a "universal" grammar as Chomsky posited.

> One fruitful avenue of research is what elements of language are shared by humankind's animal cousins. Birds can use a small number of units to make an infinite series of different calls, as humans do with words. Chimps and other apes can learn hundreds of hand signs, and even combine them in crude but creative ways. Michael Corballis, a psychologist at the University of Auckland, thinks that gesture was crucial to the rise of complex language,

a theory he expands in "The Truth about Language"?, to be published next year. Sign languages have all of the complexity of spoken ones, and deaf children even 'babble' with their hands just as hearing children do with their mouths [47].

Which are the most distinctive traits between the numerous human languages? A very interesting case is the one of anumeric people - one of the most intriguing human languages is the language that does not use numbers or any concepts related to mathematics. These so called anumeric people have developed a language without the necessity of quantifying anything. Therefore mathematics is not biologically innate to humans, regardless of how universal mathematics is.

But what is innate and common to all human languages? The answer might still lie in biology.

Genetic determination and social learning are fundamentally different processes. We are born with a genetic template that allows us to learn a language effortlessly between the ages one and four but the language we learn is completely determined by social input during this period and we learn it from others. It is part of our culture [94].

No animals have all the attributes of human minds; but almost all the attributes of human minds are found in some animal or other [24].

Animals brought up with humans or in the vicinity of humans can communicate much more and much better. But there is a difference between animals being able to communicate with humans in a way that is closer to language (such as signs and identifying objects) and the way animals communicate with their peers in the same species. There is a difference between communication that is innate and largely used and communication that is acquired and learned at the direction or design of another species.

This unsettled debate thus spurred the idea that communication that is basic is different from communication that is full of narratives, abstract concepts and also relies on grammar.

Complex ways of doing things which are passed down not by genetic inheritance or environmental pressure but by teaching, imitation and conformism have been widely assumed to be unique to people. But it is increasingly clear that other species have their own cultures, too [24].

And one of the features of language and culture that is common across all species is that it is determined by the social structure of the group.

Most cultures distinguish between outsiders and insiders and animals are no exceptions [24].

And what is different between animals and humans is the transmission of language and culture intergenerationally and the possibility of recording information for individuals outside the group.

Cultural transmission across generations, in animals, does not happen to the same extent that it happens in humans—there is no way to record the culture once the oldest animals in the tribe or group have died.

Whitehead and Rendell showed in the "Cultural Lives of Whales and Dolphins" that cultural transmission through imitation and teaching the young has had an enormous impact in the development of "vertical" culture and communication within the species, while the interactions with the environment have shaped the "horizontal" cultures [94]:

> "So the cultural transfer of information is, potentially at least, much more flexible than genetic reproduction. The products of genes change only on intergenerational timescales. In some cases, especially when a culture is conformist and learned largely from parents, it can be as stable as the product of genes; elements of Judaism, for instance, have changed little over thousands of years, but few would argue that the religion could exist without cultural transmission. At the other end of the scale, when culture is learned primarily from peers, it can be highly ephemeral, spreading fast and dying faster - think pop music or fashion. For a short time, such cultures can have immense influence on behavior and then they are gone[...]
>
> The relationship between learners and demonstrators is also important: "vertical" cultures are learned from parents, "oblique" transmission is from other members of an older generation in a form of information transfer that we have institutionalized in schools, while peers of similar age exchange "horizontal" cultures. Vertical cultures can be very stable. Religious rituals and languages, for example, can be traced back to thousands of years. In the case of language, the vertical transmission is so strong that it comes to mirror the transmission of genes from parent to child and results in linkages between genetic and linguistic diversity. In contrast, horizontal cultures – fashion and popular music styles, for instance – are faster moving and subject to very rapid change and are also generally much more transitory. In the extreme, fads can arise, spread through and then disappear from a community in a fraction of a human generation" [94].

6.3 Computational Linguistics and Communication Complexity

A large body of work exists on emergence of communication and growth of language networks. Existing research shows varying approaches and comparisons of methods to study the evolution and growth of human languages, animal signaling, biological communication, etc., focusing individually on each one of them. There have been numerous experiments and observations on all of these types of interactions which are reviewed here.

A very valuable paper by Sole et al. [84] explores various mathematical and statistical methods which can be applied to study language dynamics such as diversity, differentiation, evolution, and disappearance. Quite a few existing models are explored to understand ecological dynamics, trying to focus on analogy between linguistics and ecology. Another work, again by Sole [29] reviews the state of the art on semantic networks and their potential connection with cognitive science. The emergence of syntax through language acquisition is used as a case study to illustrate how such an approach can shed light into relevant questions concerning language organization and its evolution. Further, continuing with analysis of human communication networks, an excellent demonstration of language evolution by building a multi-agent model has been covered in the paper by Tao Gong et al. [32]. The model simulates the emergence of a compositional language from a holistic signaling system, through iterative interactions among heterogeneous agents.

These papers, however, do not investigate communication other than human language networks and how it is analogous to the other forms. This shortcoming is well addressed in Reginald Smith's [82] analysis that measures animal communication using information theory in an attempt to elaborate how closely animal communication matches, or is different from, human languages. An interesting economic model of animal communication is described in Bradbury and Vehrencamp [13] which shows the dependence of the cost and benefit of information exchange between animals. On the contrary, work by Jakob Bro-Joergensen [15] shows that animal communication cannot be viewed only as an economic process. It is argued that animal signaling depends on environmental fluctuations and signals may be repeated in spite of the cost/benefit of repetitive signaling. An influential study by Pollard [68] exemplifies this evolution of communicative complexity in animals by taking specific case study of sciurid rodents.

Apart from communication between two individuals, be it animals or humans, another type of communication which is examined is biological communication and cell signaling. The existing literature studying interactions between these biological elements points out that this form of communication has some strikingly similar characteristics to human or animal communication. In an attempt to define biological communication by T.C. Scott-Phillips [73], different approaches to have been compared to show that an adaptionist approach, which supports that response is adaptive to the signal based on cost and benefit, works better rather than just looking at it as pure non-adaptive information exchange. This illustrates that similar to human communication, biological communication takes into account the economics of information exchange. Guven et al. [36], in their study facilitated by experiments, developed a simple conceptual computational model which establishes that cell communication is influenced by external factors showing environmental learning. It is also demonstrated that degradation is necessary to reproduce the experimentally observed collective migration patterns in cells.

Humans learn from each other in unique ways—involving joint attention—that means our culture gets passed on with a particularly high-fidelity form of imitation; teaching, as well, plays a role in how we pass on culture [94].

Therefore the literature from computational linguistics and biological networks shows how interactions in human, animal and biological communication under different settings, as we also implemented in our agent based model, are taking into consideration the costs and benefits to the sender and receiver of information and how it leads to birth, evolution and death of semantic networks. The primary aim is to compare the different types of communication through simulation using the model demonstrating the exchange of memes between two networks.

Chapter 7
Applications of Language Emergence: The Future of Languages, Scenarios, and How to Use Computational Social Science Tools

7.1 Communication by Design and Its Economy

In an upcoming paper, I am proposing the details regarding a comprehensive research project on communication from cells to societies in a much broader way that I was able to only tackle through questions in this book. I am basically proposing to create an Atlas of Communication that would give us an overall picture of both the evolution of communication and the types of communication we currently know and understand [9].

If our assumptions are that life laws follow physics laws, and that there is more variation in life (biological systems) than in physical systems (cells are made of atoms; physical systems don't follow biological laws), then we can start creating some hierarchies for understanding communication evolution and scalability: communication is life-dependent and intelligent communication is civilization-dependent, which means that our best way to find general patterns of communication in the universe is to look for general patterns of communication in life and civilization on Earth and to understand the limits of variations in the laws of communication within life from those within civilizations [9].

Communication is as ubiquitous as information, yet while there is an integrated scientific discipline of information science with theories and applications, there is no integrated scientific discipline of communication science, that would study not only "local" patterns and observations of human communication, but the broader phenomenon of communication in the living world.

Communication is being researched by the vast majority of disciplines, from biology (species specific communication) to linguistics (human languages), to communication theory (context specific human communication: i.e., parent–children, political, social communication) to business and economics (by organization specific communication). This domain specific research on communication has given us very rich details about specific, local communication patterns and the role

A. Berea, *Emergence of Communication in Socio-Biological Networks*,
Computational Social Sciences, https://doi.org/10.1007/978-3-319-64565-0_7

of communication in other phenomena (i.e., predation, mating, immunotherapy, organizational functions, social movements, a.s.o.) [9].

The goal of such research would be to explore the effect of different informational structures (networks vs. symbols vs. images vs. text vs. probabilistic vs. Boolean vs. sound vs. any other form of information representation) into the evolution and emergence of communication patterns and to create a taxonomy of communication that includes both the organic and designed communication (and any mix in between) as a better representation of the physical and informational economy as a whole. Language has largely evolved organically and endogenously and with few exceptions language is essentially an informational representation of the society and culture where it was used.

More than languages, which are deeply dependent on the localized communities, art and symbols have evolved organically as efficient ways of global communication before we had emoticons and Internet (emotions and visual human behavior are perceived as such in any place on the globe). Therefore the methodological question is whether we can integrate organic, global ways to store information (qualitative visual data) with designed global ways to communicate (digital, mathematical data) in a way that can become a universal standard for storing, understanding and communicating data. The big question I would like to answer is: How will communication evolve in the near future? The applied question I would like to answer is what will be the next unifying standard in digital communication? Is this a blend of visual symbols and data science techniques?

The value of answering the question is both of a basic scientific understanding (since we haven't really bridged the evolution of signaling with the evolution of grammar and with the evolution of computer language structures), and an economic importance, as on the long run it has the possibility to advance the next innovations in terms of global technologies or global communication, as well as to give us an understanding of the rifts between the physical economies and the informational economies.

Some possible applications of this idea relate to creating universal standards for data science transfer or communication, unifying standards of new systems of communication (commercial space), or creating a universal language for human–soft AI interactions.

7.2 Natural Language Processing, Visual Communication and AI

Are metaphors visual language?

The Bayeux Tapestry that was mentioned above is famous for its length and the dynamic representation of many scenes in a narrative and metaphorical way. It is a visual, metaphorical record of an event that is dynamic, but from which it extracted the most meaningful scenes and colors and visual cues. At the same time, it was embedded in information technology available at the time (textiles) and precursor to modern computers.

As we look around a modern city or even that hunter-gatherer village, technology is the predominant manner in which culture is expressed [94]. Technology comprises the tools, techniques, and crafts that we learn. Technology allows us to construct our own ecological niches and, thus, to change our environment [94].

This book is largely based on the same ideas, that we need to revisit the original theory of information from Shannon if we want to understand meaning and semantics, which is what living organisms and particularly organisms that developed language (humans, apes, dolphins, and whales) do; unlike the living world, computers only keep on crunching data unselectively and humans will always be required to interpret and reinterpret and more software will be created for this and therefore more noise (I show that in the agent-based model) and the more noise they create, the more human supervision required to sort through the noise, and so on... therefore a "dilution" of meaningful, useful, interpretable information in the age of supercomputing and big data.

Current information models are very good at modeling syntax, metadata, and context-free structures, which is great for scalable and generalizable phenomena (such as the information and digital revolutions have shown), but the next paradigm should be to understand the "semantics" and the "meaning" of information representation in our living systems, not only in our computers, and that is very difficult to capture with 0 and 1. Otherwise we will end up drowning in the coordination and communication costs created by lots of data and we will look for solution into more software and more software will create more noise, unless this software can take on the task of interpreting, which is unlikely, given that information is just 0 and 1 (one good example is the HR software that looks through piles and piles of resumes checking only by certain keywords and thus potentially the best candidate will be left out, while selecting candidates that you still don't know if they are good fits only from the matching of the words).

What I am showing in the model is that if we pay attention to some general rules or structures that we know exist in the living world (such as awareness, biological decay, natural memory retention, and cost and benefit analysis—we know evolution of biology and informal social structures and swarm intelligence tend to be on the efficiency and positive marginal benefit side, otherwise they would not survive—therefore acting as adaptation and selection mechanisms based on the hardwired biological and social structures), and we incorporate these in the standard model of information exchange, then we can arrive at a natural, organic, evolution of communication, with more meaning and less noise.

Overall, the incomplete and messy story of language and communication evolution is so intrinsically part of our nature, that it is hard to consider it just any emergent phenomenon. It is within the fabric of our biology and sociology. Just as atoms are part of our physics, and as cells are part of our biology, communication is part of our sociality. Cells, plants, animals and humans cannot exist in isolation. We have developed various ways of communication in order to help us better evolve collectively and perhaps survive better, longer and create civilizations.

Communication is the missing link in the fundamental triad of matter-energy-information when we give information the human dimension and interpretation. If information can and may exist outside a system's power of interpretation, communication cannot exist outside that. Communication is as intrinsic to a system of communicators as are atoms to a molecule or cells to an organism.

Another major idea I am proposing is that communication has meaning, whether we are talking about communication at the cellular level, between organisms of the same species or organisms of different species. This meaning can be represented by either the chemicals and signals needed by the cell or the organism, whether it is a signal important for survival (prey-predator or for reproduction) or whether we are considering human languages. A particular case is represented by artificial communication—computers talking to each other or de-contextualized information and data sets.

But what sets biological and social systems apart from computers is the physical and memory (information storage) basis of communication that obeys certain physical decay or memory reinforcement laws that are not specific to computer or artificial systems.

Schalow argues that the more technologically developed we become and with the increased probability of artificial intelligence systems, the "balance of power" in communication will lean from the biological realm into the non-biological one [72]. The reason is that there are natural and biological barriers and very low incentives for species to communicate with other species, while the "non-biological" systems are using the same systems of signaling and communication, while also having the possibility of creating new ones, that would be incomprehensible by humans. The author has a bleak vision for the future of interspecies communication. His basic arguments are that due to the principle of mutualism, which we have employed in our model as well, there is little to no incentive for the biological systems to communicate, while the AI will turn the rules of the game in ways we simply cannot yet predict.

One of the most interesting, with broad scope and length research projects currently undergoing is the interspecies Internet, an initiative launched at a TED talk event in 2013 by the musician Peter Gabriel, one of the founders of Internet, Vint Cerf, a cognitive psychologist, Diana Reiss and physicist Neil Gershenfeld. The idea behind this project is to connect animals through technologically enabled devices, so that they could communicate with each other, we could study how they would communicate using technology and how they would communicate with us. More than this, an interspecies internet would enable us to discover patterns of communication among lifeforms that we have not been able to grasp before and this undoubtedly could lead to some general theories and patterns of communication. The interspecies internet is a very nice and challenging idea and we don't know if and when this will happen yet. But it is important, nevertheless, to recognize the importance of interspecies communication, of how much we don't know yet about it and how important this research area is for our near future.

The way we can think about communication with other entities or other species is on either of these communication avenues presented in this book: chemical communication, biological communication, technological/cultural communication and ultimately fully understood language. If we can communicate with species that share similar biological and social traits with us through body language and basically biological signals, then we can think of communicating with other technologically advanced species through similar features from the technological cultures. What about an economic means of communication? We know that trade has been instrumental in the spread of cultures and languages in our history. What aspects from the economic behavior can we consider to become features of communication?

Is the evolution of civilization predictable? Are there only a few paths or options that we can think of in terms of civilization evolution? From biology to artificial intelligence, is there only one way a civilization can emerge and expand beyond the planet where it has emerged? Does communication have anything to do with this? Communication is perhaps that linkage between the emerged intelligence and the designed intelligence that allows for a deeper understanding of our place in the universe.

For example, Lincos is one of quite a few invented, designed languages. Lincos comes from Lingua Cosmica, and was particularly designed for communication with extraterrestrial intelligence, if such intelligence exists. Lincos was developed in the 1960s as a spoken language, with sounds and phonemes, but at the same time based on basic logic and mathematics. But is Lincos that fundamental means of communication that we will all eventually learn? Somehow I doubt that.

7.3 Communication Across Time, Species and Innovative Means of Communication

One of the certainties of the future of languages is that the number of human languages currently spoken in the world is rapidly declining. As cultures become more globalized, so do languages and the natural question that comes up is whether we should preserve languages or just let them evolve? And another question is, from the currently spoken languages, should we try to preserve any them as they are or just let them evolve as well?

How much design and how much evolution? How much preservation in language and why?

Is it possible for some artifacts, such as digital communication and information, to emerge spontaneously instead of being designed by an intelligent species that can alter its evolution with technology? Is it possible for 0/1 communication to emerge naturally instead of being made by someone? Or can we consider the 0/1 information and communication part of our natural evolution, making the big, a priori assumption that technology and Moore's law follow a coevolution with us without us having much interference with it? Can we consider the advancements and the evolution of artifactual communication as part of our natural evolution, that would be reproducible under the same circumstances given by civilization and technology, or is it something else altogether?

These are some very big questions that help us at least frame a conversation about the future of communication for our civilization and perhaps give us a very tiny glimpse into general, universal patterns of communication, if such patterns exist. Under slightly different circumstances, would nature develop cellular communication that is synonymous with the artificial intelligence in our case?

Currently we don't have a very good idea or an integrated picture about the evolution of communication on our planet, from the first cells to the most advanced institutions or forms of collective organization of our civilization. This exercise would require a broad spectrum interdisciplinary research that would involve, on one hand, an understanding of communication in biology from cells to organisms, an understanding of communication in social and collective groups, from organisms to humans, an understanding of the evolution of natural language, an understanding of the evolution of collective and organization behavior and an understanding of the coevolution of communication with the one of the civilization. Given that we have a good picture about how our civilization actually communicates, we can explore potential scenarios where we relax one or more assumptions from this picture, such as what if cells on a different world do not communicate chemically, would that imply different forms of organization between higher organized intelligent species? What if organisms do not develop sight and visual communication cannot be recorded, what would that mean for the civilization? Is intelligent communication always dependent on social or collective forms of behavior or can there be entities that do not connect with peers?

Does a civilization need an economic system in order to achieve artifactual communication? Does a civilization need an economic system in order to innovate general, not culturally localized communication? Would any intelligent life have an economic system? Would they have trade and division of labor? Would they have different forms of organization? Humans have developed a diversity of forms of organization; the evolution of sociality and economic or political or entrepreneurial behavior is not reproduced or recognized in any other species. But the civilization as we know it comprises culture and social behavior that also gave us ultra-specialization in technological, artistic, economic (including financial), and religious behavior. We cannot find these in other living systems. Is this crucial for the development of a civilization, and if so, what is the relationship between the human-specific forms of behavior and communication? Would another intelligent life have to come up with similar forms of behavior and organizing before being able to come up with intelligent ways of sending or receiving communication? (see Fig. 7.1).

My intuition is that, at least on Earth, we will advance collective technologies that will span the globe and communicate universally, such as intelligent collective infrastructure systems. For example, the intelligent transportation systems of the future will use communication through symbolic or semiotic code that would be very easy to understand both the humans and the machines (aircraft, computers, radars, a.s.o.). And research shows that there is a universality in the biological

Fig. 7.1 The future of communication will have to decode the pairwise interactions of initial contact within the both species and interspecies

transportation systems (i.e., capillaries) that can inform even better the way we will have to eventually design the human transportation systems.

And another way communication will be paradigmatically transformed on Earth is through DNA computing, which has the potential to be bridged with digital computing at the molecular and cellular levels.

Therefore the mapping of the ontology and taxonomy (an Atlas) of communication on Earth would help us encompass and understand better what is intelligent civilization communication, where it is situated in the bigger picture of life and where the boundaries of what constitutes intelligent civilization lie.

My assumption is that an examination of the coevolution of the largest systems of collective, social behavior from cells to organisms that migrate to humans that built artifactual global systems would also bring us closer to an understanding of how the institutions of civilization spur communication emergence.

Chapter 8
Conclusions

The premise for writing this book has been my quest to find common features of communication in our life on Earth and to understand the role communication has played in the development of social and economic behavior across species. But more than this, while writing this book, I have discovered that if we understand much better the fundamentals of communication in life and society, we can potentially design hybrid natural-computational languages that would better fit the needs of the technological advancements in artificial intelligence (will we still be able to communicate with autonomous AI?), in biotechnology (cells communicating with digital devices and vice versa or genetic programming) as well as to get a glimpse of the future of our own species: we are moving towards fewer and fewer, but more global languages and we are trying more and more to cross the interspecies barrier in communication.

Another major idea that I have tried to render through this book is that meaningfulness or semantics are not only human constructs and are not attached to human languages alone. Meaningfulness is given in our case by the context and subjectivity we are interpreting the messages that arrive at us. Meaningfulness is one way to economize communication and attach value and selection mechanisms on information, that we would otherwise be drowning in the Tower of Babel.

While machines can talk non-stop to each other and create a lot of noise for us, there is no idea of "noise" or "meaningfulness" for brainless, non-stop communicating computers. If information has no value for computers, will the artificial intelligence of the future be capable of emerging their own set of values with respect to the information they are exchanging, and thus create their own language, in the same manner we created our natural language?

In this book I have tried to show how methodological insights in interdisciplinary approaches to complex phenomena are not unidirectional, from the natural sciences to the social sciences and economics, but they can be sometimes bidirectional and if they are successful, then the methods borrowed from the social and economic sciences not only reveal new insights into the natural sciences, but in this way

A. Berea, *Emergence of Communication in Socio-Biological Networks*,
Computational Social Sciences, https://doi.org/10.1007/978-3-319-64565-0_8

one can prove that they are robust and general for a wide spectrum of phenomena describing our blended, ecosystemic world. A world where the physical, biological, social, economic and informational are not hierarchical, but interconnected in a multitude of ways. And this is also why information is so fundamental, yet so specific; so important, yet so ubiquitous.

Appendix

The pseudo-code for an agent-based model of communication. Originally this code was developed for NetLogo [96].

initialization memes; *memes are units of meaningful language*
attributes of memes:

- network-identifier; *identifies which language network the meme belongs to*
- subnetwork; *boolean to indicate whether this meme is in the signaled subnetwork*
- subnetwork-neighbor; *boolean to indicate whether this meme is a neighbor of the signaled subnetwork*
- communicated-meme; *boolean to indicate a meme that was communicated to receiving network (a copy meme)*

initialization networks; *networks are the communicators: i.e.* O_1, O_2, O_3
to reporters:
signaler; *reports which network is currently signaling the other*
receiver; *reports which network is currently receiving information*
benefit-to-sender; *the benefit to sender of successful signaling*
cost-to-sender; *the cost to sender of successful signaling*
benefit-to-receiver; *the benefit to receiver of successful signaling*
cost-to-receiver; *the cost to receiver of successful signaling*
benefit-to-O1; *the benefit to network O1 of successful signaling*
cost-to-O1; *the cost to network O1 of successful signaling*
benefit-to-O2; *the benefit to network O2 of successful signaling*
cost-to-O2; *the cost to network O2 of successful signaling*
benefit-to-O3; *the benefit to network O2 of successful signaling*
cost-to-O3; *the cost to network O3 of successful signaling*
if $number - of - networks =$ "*two*" **then**
  $random - float < 0.5$; *there is a* $50 - 50$ *chance to determine which network is first signaler* **and**

A. Berea, *Emergence of Communication in Socio-Biological Networks*, Computational Social Sciences, https://doi.org/10.1007/978-3-319-64565-0

```
    signaler = "O_1" and
    receiver = "O_2"
else
    signaler = "O_2"
    receiver = "O_1"
end if
if number − of − networks = "three" then
    random−number < 0.33; there is a 33% chance for each network to determine
    the first signaler
    signaler = "O_1" and receiver = "O_2"
end if
if random − number < 0.66 then
    signaler = "O_2"  and
    if random − float < 0.5 then
        randomly chose the receiver and
        signaler = "O_1" and
        receiver = "O_2"
    end if
end if
to choose communication method
if communication − method = "random" then
    let random-decision
    if random − decision = 0 then
        choose send-and-receive
        if random − decision = 1 then
            choose environmental-learning
        end if
    end if
end if
if forget-weak-memes TRUE then
    max − forgotten = 100*ticks/260);
    calculate max number of forgotten memes, based on the half-life formula
end if
if strengthen-memesTRUE then
    max − strengthened = memes * e^(−t/probability);
    calculate max number of strengthened memes, based on power law; this is
    equivalent to memory retention formula
end if
communication produced reaches the maximum/plateau of the learning curve
if countmemes <= 0 then
    stop
end if
if countmemesofsubnetwork >= 1/(1+e^−t) * memesofsubnetwork then
    STOP;
```

this represents the logistic, S-shaped curve of learning
end if
to sending and receiving a signal between the two networks
$create - subnetworkofthesignalerwith <= 10memes$
signaler determines whether to continue the attempt to communicate
if $BENEFIT > COST$ for sender **then**
link subnetwork memes probabilistically into the receiver **and**
create a new meme in the receiving network
end if
if $BENEFIT > COST$ for receiver **then**
STOP **and**
switch-signaler
end if
if $COST > BENEFIT$ for sender **then**
the networks do not communicate at all **and**
switch-signaler
end if
if $COST > BENEFIT$ for receiver **then**
subnetwork DOES NOT link into receiving network **and**
all the links of copied memes die **and**
the copied memes die **and**
switch-signaler
end if
to create a subnetwork of the signaling network
get a random meme from the signaling network
if $random - meme! = nobody$ **then**
this meme now identifies as being in the subnetwork **and**
all subnetwork neighbors identify as such
end if
to calculate-sender-benefit
$potential - links(memeswith[subnetwork = true] * memeswith[network - identifier = signaler]$ **and**
$benefit - to - sender = \frac{links-sender-network}{potential-links}$
to calculate-sender-cost
$cost - to - sender = \frac{nodes-in-subnetwork}{nodes-in-network}$
to calculate-receiver-benefit
if $receiver = "O1"$ **then**
$potential - links = memeswith[communicated - meme = true] * memeswith[network - identifier = "O1"]$
end if
if $receiver = "O2"$ **then**
$potential - links = memeswith[communicated - meme = true] * memeswith[network - identifier = "O2"]$
end if
if $receiver = "O3"$ **then**

$potential - links = memeswith[communicated - meme = true] * memeswith[network - identifier = "O3"]$

else

$potential - links = 0$ **and** $benefit - to - receiver = 0$ **and** $setbenefit - to - receiver = links - receiver - network/potential - links2)$

end if

to calculate-receiver-cost

value input from user

References

1. G.A. Akerlof, The market for lemons: quality uncertainty and the market mechanism. Q. J. Econ. **84**(3), 488–500 (1970)
2. R. Axelrod, *The Complexity of Cooperation: Agent-Based Models of Competition and Collaboration* (Princeton University Press, Princeton, 1997)
3. A.-L. Barabási, *Linked: The New Science of Networks* (Perseus Publishing, Cambridge, 2003), pp. 409–410
4. M. Barbieri, Biosemiotics: a new understanding of life. Naturwissenschaften **95**(7), 577–599 (2008)
5. B.L. Bassler, M.B. Miller, Quorum sensing in bacteria. Ann. Rev. Microbiol. **55**(1), 165–99 (2001)
6. M. Bateson, S. Desire, S.E. Gartside, G.A. Wright, Agitated honeybees exhibit pessimistic cognitive biases. Curr. Biol. **21**(12), 1070–1073 (2011)
7. G. Becker, Family economics and macro behavior. Am. Econ. Rev. **78**(1), 1–13 (1988)
8. A. Berea, Trade as a premise for social complexity. J. Wash. Acad. Sci. **102**, 17–38 (2015)
9. A. Berea, How does intelligent life communicate? An atlas of communication evolution based on a unified database of evidence. Astrobiology (2018)
10. L. Boroditsky, Does language shape thought?: Mandarin and english speakers' conceptions of time. Cogn. Psychol. **43**(1), 1–22 (2001)
11. U. Birchler, M. Bütler, *Information Economics* (Routledge, 2007)
12. R.J. Brachman, On the epistemological status of semantic networks, in *Associative Networks: Representation and Use of Knowledge by Computers* (New York, Academic, 1979), pp. 3–50
13. J.W. Bradbury, S.L. Vehrencamp, Economic models of animal communication. Anim. Behav. **59**(2), 259–268 (2000)
14. S. Braman, The micro- and macroeconomics of information. Ann. Rev. Inf. Sci. Technol. (ARIST) **40**, 3–52 (2005)
15. J. Bro-Jørgensen, Dynamics of multiple signalling systems: animal communication in a world in flux. Trends Ecol. Evol. **25**(5), 292–300 (2010)
16. J.T. Cacioppo, L.C. Hawkley, Perceived social isolation and cognition. Trends Cogn. Sci. **13**(10), 447–454 (2009)
17. B. Caldwell, Some reflections on F.A. Hayek's sensory order. J. Bioecon. **6**, 239–254 (2004)
18. K. Carley, On the evolution of social and organizational networks, in *Networks in and Around Organizations*, ed. by D. Knoke, S. Andrews, volume special issue of rcscarch on the Sociology of Organizations (JAI Press, Bingley, 1999)
19. N. Chomsky, J. Piaget, *Language and Learning: The Debate Between Jean Piaget and Noam Chomsky* (Harvard University Press, Cambridge, 1980)

A. Berea, *Emergence of Communication in Socio-Biological Networks*,
Computational Social Sciences, https://doi.org/10.1007/978-3-319-64565-0

20. J.S. Coleman, E. Katz, H. Menzel, *Medical Innovation: A Diffusion Study* (Bobbs-Merrill Co., 1966)
21. L.M. Cosmides, J. Tooby, Cytoplasmic inheritance and intragenomic conflict. J. Theor. Biol. **89**(1), 83–129 (1981)
22. R. Dawkins, *River Out of Eden: A Darwinian View of Life* (Basic Books, New York, 1996)
23. P.S. Dodds, R. Muhamad, D.J. Watts, An experimental study of search in global social networks. Science **301**(5634):827–829 (2003)
24. The Economist, Animals think, therefore.... The Economist, November 2016
25. Elephant voices. https://www.elephantvoices.org/elephantcommunication.html
26. C. Emmeche, The computational notion of life. Theoria - Segunda Epocha **21**(9), 1–30 (1994)
27. P. Erdös, A. Rényi, On the evolution of random graphs. Publ. Math. Inst. Hung. Acad. Sci. **5**, 1761 (1960)
28. P. Erdös, A. Rényi, On the strength of connectedness of random graphs. Acta Math. Acad. Sci. Hung. **12**, 261–267 (1961)
29. R. Ferrer i Cancho, R.V. Solé, Least effort and the origins of scaling in human language. Proc. Natl. Acad. Sci. **100**(3), 788–791 (2003)
30. N. Friedkin, Information flow through strong and weak ties in intraorganizational social networks. Soc. Netw. **3**, 273–285 (1982)
31. A. Gomes, G. Ricardo, N. El-Hani Charbel, Q. João, Towards the emergence of meaning processes in computers from Peircean semiotics. Mind Soc. **6**(2), 173–187 (2007)
32. G. Gong, P. Rosa-Neto, F. Carbonell, Z.J. Chen, Y. He, A.C. Evans, Age- and gender-related differences in the cortical anatomical network. J. Neurosci. **29**(50), 15684–15693 (2009)
33. M. Granovetter, Strength of weak ties, a network theory revisited. Soc. Theory **1**, 201–233 (1983)
34. M. Granovetter, Economic action and social structure: the problem of embeddedness. Am. J. Soc. **91**(3), 481–510 (1985)
35. R.R. Gudwin, F.A.C. Gomide, An approach to computational semiotics, in *Proceedings of the 1997 ISAS Conference*, Gaithersburg, MD, 1997, pp. 467–470
36. C. Guven, E. Rericha, E. Ott, W. Losert, Modeling and measuring signal relay in noisy directed migration of cell groups. PLoS Comput. Biol. **9**(5), e1003041 (2013)
37. P. Hagmann, et al., Mapping the structural core of human cerebral cortex. PLoS Biology **6**(7), e159 (2008)
38. M. Hannan, J. Freeman, Structural inertia and organizational change. Am. Sociol. Rev. **49**, 149–164 (1984)
39. F.A. Hayek, *The Sensory Order: An Inquiry into the Foundations of Theoretical Psychology* (University of Chicago Press, Chicago, 1776)
40. F.A. Hayek, The use of knowledge in society. Am. Econ. Rev. **35**(4), 519–30 (1945)
41. J. Hirshleifer, Economics of information. Where are we in the theory of information? Am. Econ. Rev. **63**(2), 31–39 (1973)
42. J. Hirshleifer, A. Glazer, *Price Theory and Applications* (Prentice Hall, Englewood Cliffs, 1992)
43. J.H. Holland, *Signals and Boundaries: Building Blocks for Complex Adaptive Systems* (MIT Press, Cambridge, 2012)
44. D.W.E. Hone, D. Wood, R.J. Knell, Positive allometry for exaggerated structures in the ceratopsian dinosaur Protoceratops andrewsi supports socio-sexual signaling. *Paleontologica Electronics*, http://palaeo-electronica.org/content/2016/1369-sexual-selection-in-ceratopsia (2016)
45. M.O. Jackson, *Social and Economic Networks* (Princeton University Press, 2008)
46. M.O. Jackson, B.W. Rogers, Meeting strangers and friends of friends: how random are social networks? Am. Econ. Rev. (2007)
47. Johnson, You tell me that it's evolution? Scientists have reached no consensus on the origins of language. The Economist, 26 November 2016. https://www.economist.com/news/books-and-arts/21710783-scientists-have-reached-no-consensus-origins-language-you-tell-me-its

48. S. Kauffman, *The Origins of Order: Self Organization and Selection in Evolution* (Oxford University Press, Oxford, 1993)
49. D. Knoke, R.S. Burt, Prominence, in *Applied Network Analysis* (Sage, Beverly Hills, 1983), pp. 195–222
50. D. Krackhardt, K.M. Carley, A PCANS model of structure in organizations, in *Proceedings of the 1998 International Symposium on Command and Control Research and Technology*, pp. 113–119 (June 1998)
51. K. Kull, Biosemiotics in the twentieth century: a view from biology. Biosemiotica **127**(1–4), 385–414 (1999)
52. M. Kwiatkowska, K. Kielan, Fuzzy logic and semiotic methods in modeling of medical concepts., Fuzzy Sets Syst. **214**, 35–50 (2013)
53. L.L. LaPointe, Feral children. J. Med. Speech Lang. Pathol. **13**(1), vii+ (2005)
54. R. Lentini, N.Y. Martín, M. Forlin, L. Belmonte, J. Fontana, M. Cornella, L. Martini, S. Tamburini, W.E. Bentley, O. Jousson, S.S. Mansy, Two-way chemical communication between artificial and natural cells. ACS Cent. Sci. (2017). https://doi.org/10.1021/acscentsci.6b00330
55. S. Luca, Prehistoric signs and symbols from Transylvania (1). "the secret tablet" the neolithic and aeneolithic archaeological settlement from Tartaria-Gura Luncii (alba county). Brukenthal **10**(1), 7–16 (2015)
56. N.G. Mankiw, R. Reis, Sticky information versus sticky prices: a proposal to replace the new Keynesian Phillips curve. NBER Working Paper No. w8290 (2001)
57. S. Marcus, *Algebraic Linguistics; Analytical Models*, vol. 29 (Elsevier, New York, 1966)
58. J.H. Miller, S. Moser, Communication and coordination. Complexity **9**(5), 31–40 (2004)
59. M.L. Minsky, *Computation: Finite and Infinite Machines*. Prentice-Hall Series in Automatic Computation (Prentice-Hall, Englewood Cliffs, NJ, 1967)
60. M.E.J. Newman, The structure and function of complex networks. SIAM Rev. *45*(2):167–256 (2003)
61. M.E.J. Newman, Coauthorship networks and patterns of scientific collaboration. Proc. Natl. Acad. Sci. *101*(Suppl. 1):5200–5205 (2004)
62. V. Ostrom, *The Meaning of Democracy and the Vulnerability of Democracies: A Response to Tocqueville's Challenge* (The University of Michigan Press, Michigan, 1997)
63. E. Ostrom, Beyond markets and states: polycentric governance of complex economic systems. Nobel Prize Lecture, 2009
64. S. Partan, P. Marler, Communication goes multimodal. Science **283**(5406), 1272–1273 (1999)
65. H.H. Pattee, Cell phenomenology: the first phenomenon. Prog. Biophys. Mol. Biol. **119**(3), 461–468 (2015)
66. I.A. Paul, Enigma tablitelor de la tartaria. Conferintele Bibliotecii Astra, 2011
67. S. Pinker, P. Bloom, Natural language and natural selection. Behav. Brain Sci. **13**(4), 707–727 (1990)
68. K.A. Pollard, D.T. Blumstein, Evolving communicative complexity: insights from rodents and beyond. Phil. Trans. R. Soc. B **367**(1597), 1869–1878 (2012)
69. Y. Prat, M. Taub, Y. Yovel, Everyday bat vocalizations contain information about emitter, addressee, context, and behavior. Sci. Rep. **6**, 39419 (2016)
70. T. Riede, C.M. Eliason, E.H. Miller, F. Goller, J.A. Clarke, Coos, booms, and hoots: the evolution of closed-mouth vocal behavior in birds. Evolution **70**, 1734–1746 (2016)
71. E.M. Rogers, *Diffusion of Innovations* (Free Press, New York, 2003), p. 551
72. T. Schalow, Mutualism and knowledge sharing in an age of advanced artificial intelligence. Academic Conferences International Limited, September 2015
73. T.C. Scott-Phillips, On the correct application of animal signalling theory to human communication, in *Proceedings of the 7th International Conference on the Evolution of Language*, pp. 275–282 (2008)
74. W.A. Searcy, S. Nowicki, *The Evolution of Animal Communication: Reliability and Deception in Signaling Systems* (Princeton University Press, Princeton, 2005)

75. J.R. Searle, D.C. Dennett, D.J. Chalmers, *The Mystery of Consciousness* (New York Review of Books, New York, 1997)
76. B. Seaton, *The Language of Flowers: A History* (University of Virginia Press, Charlottesville, 2012)
77. S. Shane, S. Venkataraman, The promise of entrepreneurship as a field of research. Acad. Manag. Rev. **25**(1), 217–226 (2000)
78. C.E. Shannon, W. Weaver, *The Mathematical Theory of Communication* (University of Illinois Press, Urbana, 1949)
79. H. Simon, *The Sciences of the Artificial* (MIT, Cambridge, 1996)
80. H.A. Simon, Near decomposability and the speed of evolution. Ind. Corp. Chang. **11**(3), 587–599 (2002)
81. B. Skyrms, *Signals: Evolution, Learning, and Information* (Oxford University Press, Oxford, 2010)
82. R.D. Smith, Distinct word length frequencies: distributions and symbol entropies. arXiv preprint arXiv:1207.2334 (2012)
83. R.V. Sole, B. Corominas-Murtra, J. Fortuny, Diversity, competition, extinction: the ecophysics of language change. J. R. Soc. Interface **7**(53), 1647–1664 (2010)
84. R.V. Solé, B. Corominas-Murtra, S. Valverde, L. Steels, Language networks: their structure, function, and evolution. Complexity **15**(6), 20–26 (2010)
85. G.J. Stigler, The economics of information. J. Polit. Econ. **69**(3), 213–225 (1961)
86. J.E. Stiglitz, C. Shapiro, Equilibrium unemployment as a worker discipline device. Am. Econ. Rev. **74**(3), 433–444 (1984)
87. The National Academies Keck Futures Initiative (ed.), *Collective Behavior – From Cells to Societies*. The National Academies Keck Futures Initiative (2015)
88. R. Thom, Structural stability and morphogenesis. Topology **8**, 313–335 (1969)
89. D.A. Vakoch, A.A. Harrison, *Civilizations Beyond Earth: Extraterrestrial Life and Society* (Berghahn Books, New York, 2011)
90. T.W. Valente, Social network thresholds in the diffusion of innovations. Soc. Netw. **18**(1), 69–89 (1996)
91. R. Vigo, Complexity over uncertainty in generalized representational information theory (grit): a structure-sensitive general theory of information. Information **4**, 1–30 (2013)
92. C.H. Waddington, *The Strategy of the Genes* (George Allen and Unwin, London, 1957)
93. D.J. Watts, S.H. Strogatz, Collective dynamics of 'small-world' networks. Nature **393**, 440–442 (1998)
94. H. Whitehead, L. Rendell, *The Cultural Lives of Whales and Dolphins* (University of Chicago Press, Chicago, 2014)
95. W.A. Wickelgren, Single-trace fragility theory of memory dynamics. Mem. Cognit. **2**(4), 775–780 (1974)
96. U. Wilensky, Netlogo (1999). System available at http://ccl.northwestern.edu/netlogo
97. J.T. Wixted, The psychology and neuroscience of forgetting. Ann. Rev. Psychol. **55**, 235–269 (2004)
98. J.T. Wixted, E.B. Ebbesen, On the form of forgetting. Psychol. Sci. **2**(6), 409–415 (1991)
99. M. Wöhr, M.L. Scattoni, Behavioural methods used in rodent models of autism spectrum disorders: current standards and new developments. Behav. Brain Res. **251**, 5–17 (2013)
100. G.K. Zipf, *The Psycho-Biology of Language* (Houghton Mifflin Company, Boston, 1935)
101. K. Zuberbühler, Interspecies semantic communication in two forest primates. Proc. R. Soc. Lond. B Biol. Sci. **267**(1444), 713–718 (2000)

Index

A. Berea, *Emergence of Communication in Socio-Biological Networks*,
Computational Social Sciences, https://doi.org/10.1007/978-3-319-64565-0

Printed in the United States
By Bookmasters